Solar Revolution

Solar Revolution

The Economic Transformation of the Global Energy Industry

Travis Bradford

The MIT Press
Cambridge, Massachusetts
London, England

MIT Press books may be purchased at special quantity discounts for business or sales promotional use. For information, please e-mail <special_sales@mitpress.mit.edu> or write to Special Sales Department, The MIT Press, 55 Hayward Street, Cambridge, MA 02142.

This book was set in Sabon by SPI Publisher Services.

Printed on recycled paper and bound in the United States of America.

Library of Congress Cataloging-in-Publication Data to come

Bradford, Travis.
Solar revolution : the economic transformation of the gloabal energy industry / Travis Bradford.
p. cm.
Includes bibliographical references and index.
ISBN 978-0-262-02604-8 (hardcover: alk. paper)
ISBN 978-0-262-52494-0 (paperback)
1. Solar energy industries. 2. Solar energy—Economic aspects. 3. Solar energy—Social aspects. 4. Power resources. I. Title.

HD9681.A2B73 2006
333.792'3—dc22

2006044432

Illustration and table credits can be found on page 223.

10 9 8 7 6

To my parents—

Wayne, who taught me never to fear
honesty or hard work,

and

Susan, who taught me never to fear
anything else.

Contents

Preface: The Future of Energy

This is a book about the future of energy. Even without a deep analysis of the energy industry, most people fundamentally understand that our current energy system is ultimately unsustainable and that renewable energy (including solar energy) will be an inevitable part of our common future. Global economic, environmental, and social pressures are driving our species and our economies to change how we harness vital energy, and these pressures will intensify as we approach the middle of the twenty-first century and expand to an estimated population of ten billion inhabitants on the planet.

Many of the greatest hurdles we will face in the next fifty years will be a direct result of how we currently and eventually decide to procure the energy necessary to sustain our lives and our standard of living. Human-induced climate change, resource wars over energy supplies, and cycles of deforestation, famine, and poverty that result from our insatiable appetite for energy are not new problems. Humans have grappled with these problems for centuries. The difference today is that these problems have accelerated in scale and potential repercussions to global proportions.

Inevitably, the threats that our relationship to energy creates will be mitigated when motivation and opportunity collide. This could happen when businesses and government compensate for the risks and costs of our current energy system with effective foresight and coordinated planning or, alternatively, when we are forced to change in response to a 1970s-style energy crisis. Whatever the catalyst, the industrialized and developing nations of the world will eventually address these issues by using energy more efficiently and by developing and deploying local, sustainable, renewable energy sources.

Many such energy-generation solutions are being pursued, including nuclear power and renewable wind, biomass, and geothermal energies. Businesses and policy makers are currently pursuing choices based on their respective natural-resource endowments, technical expertise, and political will. For example, Iceland is tapping into its vast stores of geothermal and hydroelectric energy in an attempt to become the world's first fossil fuel–free economy. The countries of northern and western Europe (including the United Kingdom, Denmark, and Germany) are taking advantage of their ample wind resources to lead the world in wind-power deployment. Land-rich but oil-poor Brazil is deploying biofuels to power its transportation infrastructure at a lower cost than traditional gasoline or diesel fuel. Each of these developing energy sources has a role to play worldwide, and many will be components of the solutions that are ultimately employed.

Various solar-energy-generation technologies—including direct electricity generation from photovoltaic (PV) cells—also continue to be researched and deployed. Although PV technology is conceptually simple—harnessing the sun's energy on a solid-state device—generating electricity with PV cells is generally assumed to be both too expensive and too far behind in terms of market penetration to have a meaningful impact on the juggernaut of the world energy infrastructure. Partially because of solar energy's false promises in the 1970s, the technology is widely seen as a desirable but uncompetitive energy source in all but niche markets and remote small-scale power applications. However, developments in the PV industry over the last ten years have quietly transformed solar energy into a cost-effective and viable energy solution today.

In many markets such as Japan, Germany, and the American Southwest, PV electricity has already become the energy choice of hundreds of thousands of users. From this established base, the technology of PV is poised to transform the energy landscape within the next decade as relative prices of this technology versus existing sources make it increasingly competitive. PV technology's relative cost-effectiveness when compared to traditional energy choices and even many of the "new renewables" such as geothermal, wind, biomass, or ocean power will ensure its continued market penetration. Although it will be many years before solar energy provides a substantial amount of the world's energy generation, awareness of the inevitability of the solar solution will have a surprisingly dramatic impact on electric utilities, government policy makers, and end users much sooner than most predict.

At its root, the shift to solar energy will be due to two complementary economic drivers in the energy industry that affect the configuration of energy supply and demand. The first driver relates to *what types of energy source are used* to power modern industrialized and developing economies. Pressure to develop sources of clean, renewable energy is growing because of the increasing costs and risks of securing traditional energy supplies, the increasing need for more energy as countries like China and India industrialize, and a growing understanding of the environmental effects of traditional sources of energy.

The second driver relates to *how and where energy is being generated*. Over the next few decades, industrial economies will shift away from large, centralized energy production toward smaller, distributed energy generators, primarily because end users will increasingly have cost-effective options to avoid the embedded costs of the existing energy infrastructure. This trend toward distributed energy is also true for the billions of people who live in developing economies (where most of the global growth in energy use is projected to occur) and who do not currently have access to large, centralized electricity grids and distribution systems. As these two drivers combine to change the economics of energy, much of the world will find it economic to use locally generated, clean, renewable energy. This book discusses the inevitable conclusion of these two trends—when, where, and why they will occur.

The research that led to this book did not begin with the supposition that such a clear energy path existed. It began with the broader question of where the natural momentum of the global energy industry has been leading and what trends would determine its future. The inevitabilities regarding solar energy became apparent only through an understanding of the natural economic forces that were transforming the industry, the changing relative costs and risks inherent in the various energy technologies, and the surprisingly close proximity of transition points for various energy users that would alter their decision making. But while inevitability alone is an interesting concept, it is not particularly useful without the answers to three pivotal questions: when will this inevitability arise, what challenges stand in the way between today's status quo and the inevitable configuration, and is such inevitability desirable enough that efforts should be made to accelerate it?

To answer these questions, this book examines the entire energy cycle that dictates our relationship through energy to other people and to the planet rather than just the energy infrastructure that utility providers and fossil-fuel suppliers typically describe. Only by placing global energy use in this greater context can we properly evaluate the decisions that we as individuals and as a society will ultimately make. In determining which energy options will prevail, a reasonable analysis must look beyond preconceptions about which one "should" succeed or which one would be "the best" solution for society. Such analysis relies too much on wishful thinking amid disparate and conflicting political and economic agendas. Instead, responsible analysis should determine how, in the course of day-to-day life and trillions of individual uncoordinated decisions, energy solutions will unfold naturally.

Forecasts of this nature are always risky. However, constructing models of the future is critical for sound decision making on important topics, and various forecasting approaches can be applied. Some people build mathematical models, some use broad philosophy, and still others take a business approach. The forecasts herein use a combination of economic and business modeling because, in the end, the relevant question is how the global energy industry and its economic agents will behave. In business, when managers are attempting to forecast market conditions over long periods of time, specific forecasts are not always possible or even useful. Understanding and predicting key market drivers and the ways that they will change over time are how the underlying tectonic, and eventually determinative, forces are detected. Correctly assessing these key drivers and using them to economic advantage is what separates highly successful businesspeople from the pack. When the key drivers in the global energy industry are identified, they expose the fallacy of the conventional logic that states that solar power is destined to be a marginal player in our energy future.

The inevitability of solar power itself is a powerful concept, and a clear vision of the inevitable will help guide decision making today and in the years ahead. Although the size of the existing energy infrastructure and the long life of the assets employed may mean that it will be many years before the world is dominated by clean, virtually unlimited solar energy, the increasing momentum in that direction will transform the world and our expectations long before. In the end, perhaps that is the only change that is needed. It may be sufficient for now to realize that alternative paths do exist, that the goals of promoting business and the environment need not be mutually exclusive, and that progress toward a practical, sustainable relationship with our planet is not only achievable but inevitable.

Acknowledgments

This book is yet another testament to the fact that sole authorship is a team effort that is made possible by the devoted attention given to it by many people. Foremost, the members of and advisers to the Prometheus Institute deserve praise for meeting the timetables and enduring the chaos of the process; these include Greeley O'Connor, Hilary Flynn, Varda Lief, Lisa George, Pratibha Shrivastava, Suparna Kadam, Larry Gilman, and Hari Arisetty. I am particularly indebted to my agent, Sorche Fairbank, as well as Clay Morgan of MIT Press, for their vision and support of the ideas in this book.

Thanks to my friends and members of my family who have allowed me countless hours of talking through the ideas of this book. Allison Cripps, Olaf Gudmundsson, Sam White, and so many more have generously given me the time and space to work. Also, thanks to those who took the time to review versions of the manuscript—Andrew Jackson, Alex Lewin, Gagan Singh, Per Olsson, and especially Anu, for her understanding and patience throughout the process as well.

I want to thank all of those people who have given help, support, and encouragement during the writing of this book, many of whom have reviewed drafts and provided valuable insights on the history and forces shaping the evolving solar-energy industry—specifically, John Holdren, Tom Starrs, Craig Stevens, Paul Maycock, Scott Sklar, Michael Rogol, Steven Strong, Bob Shaw, Ken Lockin, Denis Hayes, Clark Abt, Hermann Scheer, John Perlin, Josh Green, and Janet Sawin. A very special thank-you goes to Jigar Shah, who has continued to be all of the above—a source of information and insight, a reviewer, and a provider of constant encouragement.

Finally, I want to thank the people who inspired me to write the book in the first place: Marina Cohen, who unwittingly provided the spark;

the authors of the hundreds of books and articles that allowed me to gain a deeper understanding of the problems we face and their potential solutions; the generations of inventors, entrepreneurs, and advocates—so many without credit—who came before us and paved the way for the transition that the world is about to experience; and finally those who read this book and become inspired to contribute to its realization.

Though a team effort by many, the book's errors and oversights remain mine alone.

I

The Inevitability of Solar Energy

1

A New Path on the Horizon

Energy is hot again. Not since the oil-price shocks of the 1970s has there been such a buzz about energy or its impact on the world economy. Newspapers and news programs increasingly focus on the issues surrounding the world's energy needs and the consequences of current global production and consumption patterns. Yet this crescendo of media stories and reports issued by the United Nations, nongovernmental organizations (NGOs), and policy think tanks has not been able to convince people and businesses that viable alternative solutions or pathways yet exist.

A growing number of environmentalists, scientists, economists, policy experts, and citizens understand that current energy dynamics dictate that the world will soon run short of relatively cheap, easily accessible oil—to be followed quickly by natural gas and coal—and that energy alternatives must be developed quickly. Because it is impossible to predict all of the variables that will drive these future changes, the consequences of delaying development of energy alternatives can be discussed only in terms of a range of possible scenarios. According to the best scenario, industrial economies will see sagging economic output and productivity and massive wealth transfers to the oil-rich countries of the Middle East by the mid-twenty-first century. The worst scenario includes global ecological melt-down and human suffering on an unimaginable scale.

However, deploying sufficient energy alternatives to help us to avoid economic and environmental crisis is a massive and daunting task that is made more difficult by inertia. Today more than 6 billion people make daily decisions about what to eat, what to wear, or what to drive. When they decide between the immediacy of a household budget and an uncertain energy and ecological future, most people frankly do not understand and cannot afford to care about the long-term impacts of their decisions.

Globally, most people lack the necessary information or day-to-day economic security that would allow them to understand and act on the long-term effects of small, daily choices. Their priorities are feeding themselves and their families, staying warm and safe, and carving out whatever security they can. To meet their vital needs, people and the societies they comprise will continue to absorb trees, fossil fuels, and food stocks unless and until accessible and cost-effective energy choices exist. The history of our species, not unlike the history of algae blooms, is a repetition of this story. It is the story of a species trying to improve its lot and, through ingenuity or chance, tapping into a new source of food or energy. This species eats and multiplies until the available food and energy are dwarfed by the population, followed by a painful adjustment in lives and economic livelihood until a new equilibrium is found. This pattern, described most eloquently by the English economist Thomas Malthus in 1798,[1] has been repeated many times in human affairs from ancient Babylon to the Roman empire to imperial China to modern Africa.

Throughout history, human beings have cleverly harnessed available energy sources in the environment by adapting sources of stored energy—first wood, then animal power, then agriculture, and finally the miracle of fossil fuels. As Malthus predicted, this improvement in our standard of living has led to a corresponding increase in human population to unprecedented levels. Historically, when resources became scarce or depleted in one geographic area, humans adapted and migrated to other areas where resources remained, often involving a costly or painful transition. In the last round of population expansion, the industrial age of fossil fuels of the nineteenth and twentieth centuries, the human race finally managed to come full circle around the globe. We have extended our reach to nearly every useful location, populated nearly every worthwhile parcel of land, and are steadily depleting the remaining energy resources. There is nowhere left to run.

Today, leaders of the industrialized world are again facing rising threats from volatile energy prices, adequate access to fuel supplies, and insecurity arising from potential nuclear states. There are new larger threats this time around, however. Since the early 1980s, a growing awareness of the causes and effects of global climate change, of the risks of resource peaking in oil and natural gas reserves, and of an aging energy infrastructure has added to the urgency of the problem. Growth in energy demand is projected to continue unabated, with much of the growth expected to occur in the burgeoning industrial societies of China, India, and other countries of the developing world.[2] Growing global demand, perpetually risky supply, and volatile prices are leading to a potential "perfect storm" of threats that weigh heavily on governments, businesses, and consumers throughout the world.

A Bankrupt Energy System

Fossil fuels in the earth's crust are a result of millions of years of layer upon layer of the detritus from oceans, swamps, forests, and ecosystems that accumulated and then was covered over to slowly transform into what we now mine or drill in the form of coal, oil, and natural gas. The vast energy latent in these substances is both portable and easily harnessed, which has allowed for the development of an industrial society based on energy-intensive devices ranging from microchips to street lights, laptops to supertankers, and V-8 rockets to 747 airplanes. Fossil fuels have enabled societies to extend life and reduce suffering but also to wage war on ever more devastating scales. Modern economies have avoided many of the pitfalls of unconstrained growth because they have been able to switch among fuel sources as necessary or useful or because they have tapped into new sources of energy to facilitate technological and economic expansion. By implicitly relying on continuing technology and productivity growth to outpace population growth and energy demand, trained economists have declared for two centuries that the theories of Malthus are "dead." Changing energy dynamics may yet prove that view overly optimistic.

To understand modern society's relationship to energy, it is helpful to think of energy as money, with corresponding categories of income, savings, and expenditures. The world's annual energy *income* is all the energy captured each year from new sources. Trees and other plants collect energy income from the sun, as do renewable-energy technologies like hydro, solar, and wind, either directly or indirectly. Renewables are renewable because they draw primarily on the earth's solar paycheck, as long as the sun shines. Yet energy income effectively shrinks if the ability to capture energy is diminished. This happens when forests are cut down faster than they can grow back and arable soils are allowed to wash away, limiting the amount of energy capture available to farmers.

The world's energy *savings* consist of all the energy that is stored, in whatever form, in various reservoirs. These reservoirs include standing forests, the thermal energy in large bodies of water, and the earth's vast (but mostly inaccessible) inner heat, uranium, and other fissionable metals. Our most accessible energy savings include the millions of years of solar energy stored in the form of fossil fuels. The energy savings of the earth, especially in its fossil fuels, are vast but finite. Despite claims to the opposite, fossil-fuel energy is not likely to be totally exhausted under any future scenario. Some amount of fossil fuel will always be available at some level of processing and at some cost. However, the looming threat of energy depletion is not about total fossil-fuel exhaustion but rather is about its impending scarcity and the resulting effects on price and availability. As the point of global peak production of fossil fuels—primarily oil and natural gas—is passed within the next decade, the vital fuels on which our global economy is founded will rapidly get more expensive. The repercussions will reverberate throughout the entire industrial infrastructure. The fact that some coal or oil is left in the ground will not be economically meaningful if the global cost for extracting useful industrial energy becomes increasingly expensive.

Energy *expenditure* in this context is simply the sum of all energy used within the global economy. The world's rate of energy expenditure is a function of global population and the average energy used by each person, and under nearly all scenarios it is expected to increase. With global population expected to reach almost 10 billion by midcentury and China and India (together comprising over a third of the world's population) expected to increase their per capita energy usage by three to five times over the next thirty years, it appears certain that global energy expenditure will continue to grow for the rest of the twenty-first century. Technology advancements will improve our ability to extract the remaining energy stored in the earth's reservoirs economically and use it productively, just as technology has done since the beginning of the industrial revolution. Unfortunately, the increasing demand for global energy expenditures continues to surpass even technology's ability to compensate.[3]

With global energy savings falling and energy expenditures rising, our global energy system is facing a real risk of bankruptcy. The global economy is drawing more rapidly on its diminishing stores of trees, soil, fossil fuels, and everything else, and its unrestrained growth in energy demand is becoming unsustainable using current energy solutions. With its global energy expenses unlikely to decrease and cheap and accessible energy stores equally unlikely to be discovered, the world must increase its energy income by finding renewable alternatives that can meet its vast needs. To be effective, such a solution—or blend of solutions—must offer a source of energy that is widely accessible and that will be chosen naturally as people obey the logic of short-term self-interest, cost, and efficiency.

Back to the Basics

For reasons that this book explores in full, solar energy will inevitably become the most economic solution for most energy applications and the only viable energy option for many throughout the world. Currently, sunlight is the only renewable-energy source that is ubiquitous enough to serve as the foundation of a global energy economy in all of the locations where energy will be required, from the industrialized world to the developing one. The evolving economics of energy reveals that electricity from solar sources has certain projected cost advantages compared to other forms of generating electricity that ensure its major role in meeting the world's energy challenge. Looking at the gap between the amount of direct solar energy being harnessed today and the amount of energy that will be required to meet increasing energy demand and replace dwindling fossil-fuel sources over the next fifty years hints at the likelihood for unprecedented growth in the solar-energy industry.

Obviously, the world will never be powered entirely by direct solar sources. Energy will always be supplied by a portfolio of technologies, including those traditionally harnessed from fossil fuels. Increasingly and dramatically over the next few decades, however, consumers will turn directly to the sun for their energy. This will happen not because solar power is clean and green but because basic economic and political reasons compel us to make this choice. At the point that the out-of-pocket real cash cost of solar electricity drops below the costs of current conventional energy alternatives (a situation already occurring in the Japanese residential electricity market), the adoption speed of solar energy will rival nearly every technological leap in history, even the rapid and transformative adoption of computers, information technology, and telecommunications in the late twentieth century. Eventually, solar energy will become a major portion of the electricity infrastructure (both the utility grid and local distributed generation) and contribute substantially to energy used in the transportation infrastructure.

Many people in government, economics, and ecology might initially find this claim difficult to accept. Conventional thought is dominated by the view that solar energy is still a long way from being cost-effective or efficient and will be doomed for decades to play catch-up with cheaper alternatives such as wind, nuclear, and biomass energies. But these assumptions rely on the traditional framework of energy cost analysis and embedded assumptions about the future that are derived by extrapolating historical trends incorrectly. Such analyses are examined in detail later in this book and shown to be incorrect and incomplete.

Understanding the nature of this transformation toward dramatically increased use of solar energy requires clear definitions of the terms of the discussion. As a first step, let us consider electricity rather than energy, which is a much broader category. Though electricity consumes roughly one third of the primary energy used in the world, it plays a fundamental role in the productivity of industrial economies and provides a vehicle for addressing similar energy issues in other energy sectors, such as transportation and heating applications.[4]

Electricity is a particularly pure and versatile form of energy. It runs computers, lights, transportation systems, and factories. Economies depend on the quality, reliability, and quantity of electricity available to them and the efficiency with which it is used. Electricity's contributions to the modern world currently rely on the large-scale electricity-distribution systems that were begun around the turn of the twentieth century by inventors and entrepreneurs like Thomas Edison and George Westinghouse. The electricity grid, through which all modern economies are powered, was dubbed the greatest invention of the twentieth century by the National Academy of Engineering in 2000, surpassing even the automobile, the airplane, and the computer in importance.[5] Over the last one hundred years, the cost-effectiveness, versatility, and reliability inherent in this grid technology led to an increase in wealth and productivity that rapidly brought lights and other appliances to many homes and businesses around the world. The grid could deliver energy to the user more cheaply—and in a far more versatile form—than coal or other forms of fuel hauled to each user's location and consumed on site. Electricity generation became less expensive over time because of the economies of scale made possible by centralized generating plants. For over a hundred years, industrialized nations have relied increasingly on the grid to supply all but the largest industrial energy users (who sometimes generate their own electricity) and in doing so have reaped substantial benefits in reduced energy costs and increased reliability of the energy sources necessary to promote industrial development.

Today—thanks to the sheer size of and number of people connected to the electricity infrastructure, a century of accumulated technical experience, and substantial government subsidies—retail prices for electricity are at their lowest levels ever. The United States, for example, has some of the lowest electricity prices in the industrial world. The cost of its residential electricity averages around nine or ten cents per kilowatt hour (kWh), the standard measurement for electricity usage and flow.[6] In other industrial countries, electricity costs vary due to differences in the mix of fuels used, fewer economies of scale, and lower government subsidies. In Japan, for instance, the retail price of electricity is about twentyone cents per kWh, while in Germany it is twenty cents per kWh.[7] These average prices can be misleading, however, and can distort analysis because local electricity prices can vary widely within a country depending on local economic factors and the type of power being generated.

A useful way to begin exploring electricity economics is to break its cost into two pieces—the cost to make the electricity (*generation*) and the cost to get it from the point of generation to where it is needed (*delivery*).[8] Different places and producers have different cost structures, but a basic rule of thumb is that residential electricity costs divide more or less evenly between generation and delivery. Using the U.S. example of nine cents per kWh, this results in 4.5 cents per kWh for the fuel and plants to make the electricity and another 4.5 cents per kWh for the grid to transmit it.[9]

These numbers represent the actual cost paid by consumers of electric power but are not *fully loaded,* using the term of accountants and economists. Fully loaded costs include the costs that are sometimes transferred to and paid by outside parties, such as costs of subsidies or pollution control. Fully loaded costs also consider the total cost of replacing the industry's capital base (including its power plants, infrastructure, and equipment), which depreciates or deteriorates every year. Worldwide, current electricity prices do not fully account for these costs, and if they did, retail electricity prices would be substantially higher. One reason they do not is that governments often take on some of the costs of building, financing, and protecting the energy business and pass on those costs to consumers in the form of taxation rather than in the cost of delivered power.

Another reason that prices do not reflect costs is that since the wave of deregulation in many industrial electricity markets in the 1980s and 1990s, newly privatized and deregulated utilities around the world have relied on the existing installed infrastructure and have underinvested in maintaining the electricity grid. The consequences, as the last few years have shown, are increasingly frequent and dramatic blackouts and brownouts and eventually will be higher costs to consumers and utilities as additional capacity is added to adequately replace this aging infrastructure. In the end, though, consumers and businesses make decisions based on out-of-pocket costs, not those that society must bear. Independent of the policies that allow these costs to be less than fully loaded, any analysis of the future of the energy industry must recognize the economic reality that out-of-pocket costs are the relevant factors. A responsible analysis of electricity-industry economics must assume that the cash price that energy users pay is the primary variable that users consider as they make decisions.

Historically, most analyses of electricity economics have looked at costs from the utility's vantage point, primarily because utilities in industrial economies currently generate well over 90 percent of all electricity.[10] Since delivery cost is essentially fixed for grid-based electricity regardless of the utility's method of generation, the standard approach to comparing the economics of various electricity sources has traditionally focused on differences in generation costs. Under this methodology, each new technology or new installation of an existing technology must show that it can generate electricity more cheaply than the installed base of electricity generators. From this perspective, only the established technologies of coal, oil, natural gas, hydropower, and nuclear energies had any hope of being economically competitive because alternative-energy technologies, with their limited scale, were perceived as too expensive or too risky to be considered by the large utility companies. As a result, economies of scale in the traditional technologies continued to be reinforced. The landscape has begun to change in the last decade, however, as a new breed of alternatives has reached the level of technical sophistication and cost to require a fundamental reexamination of electricity economics.

A Portfolio of Alternatives to Choose From

Some of the greatest optimism in the field of renewable energy has come in the last twenty years as the cost of generating utility-scale electricity through cleaner and more efficient wind power has dropped by a factor of five.[11] Today wind power, using the largest windmills at the best locations, is cost-competitive with electricity generated by many fossil-fuel plants. Globally, 6 percent of the electricity-generation capacity installed during 2004 was wind-based, and the wind-power industry is growing at more than 20 percent annually worldwide.[12] While the developments in wind power are both encouraging and exciting, wind power has limitations in its ability to supplant the bulk power needs of today's industrial economies mostly because wind is inherently unpredictable and a limited number locations have sufficient wind resources. In addition, resistance by many local communities to having wind farms in residents' line of sight has slowed the rate of adoption of wind power even when the economics are compelling.

With uneven global distribution of fossil-fuel resources and few economical, renewable resources for utility-scale electricity, nuclear power is also being revived as a potential source of electricity generation. Propelled by intense technological optimism and large government subsidies, nuclear power climbed from 2 percent of world electricity supply in 1971 before leveling off to nearly 17 percent in 1988.[13] Even before the headline-grabbing accidents at Three Mile Island (1979) and Chernobyl (1986), nuclear plant orders had dried up in the United States—the world's largest generator of nuclear power—based on the high cost of electricity generated by nuclear power.[14] While advocates of nuclear power argue that it could be made cheaper, safer, and cleaner, no credible plans are in place to accomplish any of these objectives. Nuclear-waste repositories are hotly contested as are reprocessing facilities. Some industrial nations, such as Germany, have committed to the reduction and elimination of their nuclear-power capacity. However, other governments, such as South Korea and France, are extending the life of their existing nuclear facilities and considering a revival of nuclear-power-plant construction despite the risks it may pose to the environment and global security.

Hydropower is another potential solution for global energy needs, but there are not enough commercially viable hydropower opportunities to meet rising global demand. While most industrialized nations have already developed their economic hydroelectric opportunities, many developing nations are increasingly relying on hydropower projects to meet domestic energy demand. One example is China's Three Gorges Dam, which will create a reservoir nearly 400 miles long and will displace 1.2 million people when it is finished filling in 2009.[15] Even where remaining hydropower resources can be harnessed, scientists are increasingly recognizing that the costs to the environment and the communities that are displaced by these projects are more severe than previously understood.[16]

Regardless of which traditional or alternative electricity energy technology is being evaluated, the standard operating procedure of comparing only generation costs represents an incomplete and therefore inaccurate analysis. Traditional analyses, performed from a utility's vantage point, assume that all electricity technologies rely on the electricity grid to deliver their power to homes and businesses. To assume otherwise would assume away the utility's own future in electricity delivery. Industry analysts, governments, and NGOs, either because of inertia or tacit agreement, continue to use the same assumptions and analytic tools. However, this analysis neglects the understanding that electricity users desire only to receive reliable power at the lowest cost and effort; whether they do so through the grid or not is irrelevant to them, other things being equal. If an energy source can bypass the traditional infrastructure and delivery system, delivering its power directly to the end user, then methods of comparing costs among them must reflect this change. Although solar electricity can be generated centrally and distributed over the grid, more cost-effectively than commonly appreciated, it need not be. Solar electricity can be generated almost as cheaply and easily on an individual rooftop (known as *on-site distributed generation*) as it can be at a huge, utility-operated solar-panel farm.[17] Ultimately, this new competitive landscape will change the underlying economics of energy for both centralized and distributed users.

The relative costs of solar energy and the grid electricity it replaces continue to change as well. Solar module costs per installed watt have been declining for the last decade at 5 to 6 percent per year because of technological advances, scale of production, and experiential learning.[18] Today, solar-electricity generation has reached a point of cost equivalence for millions of households worldwide and will decline even more as global solar production continues its historical growth rate of 29 percent annually.[19] The transition in solar economics is happening first in applications and in places where three factors combine—ample sun, expensive grid-based electricity, and available government incentives. For all types of users, the cost-effectiveness of solar electricity is likely to increase faster than even the most aggressive ability to increase solar-panel supply, setting up a decades-long growth scenario for this industry.

New Choices Create New Economics

Though it will be some time before solar electricity is competitive with the centralized utility-scale generators of hydro, coal, and nuclear power that run constantly, solar is already competitive with a large part of the energy-generation infrastructure that is used only during high-priced, high-demand hours. One of solar power's great attractions for utilities—apart from zero fuel costs and low maintenance requirements—is that consumer electricity demand and the power that utilities must provide throughout a typical day neatly track the daily and seasonal energy cycle from the sun. The times when energy demand is the highest coincides with those when the sun shines more brightly, including part of the electricity demand that is directly tied to the sun's availability, such as summer air conditioning.

Utilities call the electricity needed to meet this part-time demand *intermediate-load electricity*, as opposed to the *base-load electricity* that is needed twenty-four hours a day. Intermediate-load electricity is relatively expensive to generate because it comes from generators that, by definition, are used only for a portion of the day, making the electricity they generate more expensive as the cost of the generator is spread over less output. By its nature, solar power provides intermediate-load electricity. To be economic for utilities, therefore, solar-power technology needs to become a competitive producer of intermediate-load electricity, which represents 30 to 50 percent of total electric demand and is disproportionately supplied today by natural-gas generators.[20] Utilities are also beginning to realize that installing intermediate-load solar generators on the consumer side of the grid can offset the cost of upgrading transmission lines and equipment in many instances.

But utilities are not the only potential adopters of solar electricity generation. Today, distributed end users (including home and business owners) can elect to generate their own electricity with PV, but they will do so when installing solar generators on their side of the electricity grid, on a home or commercial building, becomes less expensive than buying electricity through the grid. This decision point is not hypothetical. Millions of households worldwide that are not currently connected to any grid (or are connected to an unreliable grid) find PV electricity the most cost-effective electricity solution because it represents the only viable form of modern energy available to them. More importantly, many grid-connected homes worldwide (particularly in Japan and Germany) have already elected this option through *grid-connected PV systems*. Grid-connecting a PV system eliminates the need to store daytime power for nighttime use, overcoming the inherent limitation that solar electricity generates electricity only during daylight hours. Grid-tied solar electricity is generated when the sun is shining, and the excess is stored by sending it back into the utility grid supply. At night, users purchase conventionally generated power from the grid as needed. The grid itself functions as a huge storage battery that is available for backup power and eliminates the need for system owners to install expensive equipment to provide storage and backup electricity services.

For both utilities and end users, the economic rationale for making the switch to grid-connected solar electricity will be reached in different markets with different applications at different times. Generally, though, this book shows that the transition to solar energy and electricity technology will happen much faster than most people imagine, faster even than most experts commonly predict. This transition will occur not because well-meaning governments force solar panels on reluctant markets to capture environmental benefits (although such efforts would help to accelerate the rate of global PV adoption) but rather because solar power will increasingly be the cheapest way to do what people want to do anyway—light spaces, manufacture goods, cook, travel, compute, and watch TV.

Even with solar power's current low market penetration and consequent lack of economies of scale, it is rapidly crossing over into cost-effectiveness in certain major markets. As its world market share in the energy mix climbs from less than 1 percent of new annual electricity-generating capacity and less than .05 percent of total electricity generated to hundreds of times its current level over the next half century, it will progress along its experience curve to become significantly less expensive.[21] Solar installation will occur increasingly at the time of construction for sites and buildings, which reduces the cost of installing these systems from today's primarily retro-fit installations through the efficient use of installation labor and the offset roofing and glass that PV systems replace. In addition, with so much of the cost of PV electricity in the up-front cost of the systems, improvements in financing (including wrapping PV systems into the standard mortgages of home and office buildings) will dramatically improve PV economics from today's levels. In the end, the real cost of capital to finance distributed PV systems in this way will be far cheaper than that available to utilities or any other centralized generator.

Solar electricity provides other economic advantages beyond cost-effectiveness that are important but often difficult to quantify. Two of the most important are modularity and simplicity. Thanks to modularity, solar-cell installations can be precisely sized to any given application simply by installing only as many panels as are needed. Large solar installations can be brought on-line in stages, panel by panel, unlike large conventional power plants that generate no electricity during the many years they take to build.[22] Solar panels can be serviced piecemeal, too, while the remaining panels in the array continue to make electricity uninterrupted. Solar power's physical simplicity means low training costs for users, while solar's lack of moving parts translates into high reliability and low maintenance. Long module life, on average thirty years or more, also adds to the inherent cost advantage of solar cells. As the economic playing field levels, market choices in electricity will increasingly be driven by these types of inherent advantages.

Beyond Wishful Thinking

The conclusion of the economic inevitability of solar energy has thus far been based on the assumption of improving relative economics for solar electricity. What has not been assumed is also important to consider.

The analysis supporting these conclusions does not assume that governments will do more to encourage investment in renewable energy or that governments will impose disincentives on the use of fossil fuels or nuclear power. Some governments—including those of Japan, Germany, Australia, and many U.S. states—are already promoting solar electricity by offering incentives and streamlining connections to the electricity grid. However, forecasts based on government programs that do not yet exist are irresponsible, and waiting for such programs to materialize is even more so. Many people both inside and outside government are promoting renewable energy, but the belief that a renewable-energy economy will not happen without greater government support—as environmentalists too often argue—is wrong. The shift will happen in years rather than decades and will occur because of fundamental economics.

The conclusions of this analysis do not rest on an assumed significant increase in the price of fossil fuels, though that is the most likely scenario. Few people believe that fossil-fuel costs will drop in the years to come. Indeed, many experts are predicting increased market volatility and prices, and some even predict a spike in oil and natural-gas prices to levels beyond those of the oil shocks of the 1970s, based on dwindling reserves, rising demand, low investment in supply infrastructure, and potential political instability in the largest oil-producing regions of the world (that is, the Middle East and nations such as Venezuela). The effect of such price spikes could be even more devastating to the world economy now than in the 1970s since this time the supply constraints would likely be physical and permanent unlike the artificial ones set by the Organization of Petroleum-Exporting Countries (OPEC) thirty years ago.

Technology breakthroughs are also not assumed (or required) in this analysis. What is required is continued growth in cost-effectiveness and the technical expertise of existing PV technology at recent historical rates. These improvements can easily be realized by increasing economies of scale as production continues to grow annually by double-digit percentages and as progress continues along the usual experience curve for new technologies. Both of these natural results of processes are already under way.[23] This is not to say, however, that breakthroughs will not occur. Should one of the many public or private research laboratories around the world researching photovoltaic technology make a breakthrough (for example, halving the material cost or doubling the efficiency of today's most cost-effective solar-cell design), the transition to a solar economy would further accelerate.

Numerous indirect social benefits to a transition to a solar economy are worth mentioning, even though they are not used in this book's analysis of renewable-energy economics. These indirect benefits will be substantial for every fossil-fuel-poor country in the world, from sub-Saharan Africa to most industrialized regions. Through worldwide economic growth, the switch to solar power will improve energy security and balance of trade, deliver massive direct-wealth creation to less developed countries that are solar-rich but infrastructure-poor, and create indirect wealth effects for their trading partners. The transition to solar would also limit pollution and lessen the risks posed by global climate change by reducing greenhouse-gas emissions over today's fossil-fuel-based energy sources. In addition, cheaper local energy sources would help accelerate the transition to electric- or hydrogen-powered vehicles. Wide deployment of inexpensive distributed energy would help reduce the cost of drinking water through desalination and provide cheaper water and fertilizer for agriculture. All these changes are crucial to sustaining 9 to 10 billion people on the planet by the middle of the twenty-first century.

While these social benefits are worth noting, none have been assumed because they are not necessary to the conclusion that a transition to direct solar energy is inevitable. As mentioned earlier, energy consumers—who ultimately drive economics—usually make decisions based on immediate concerns such as cash in versus cash out. To assume that such decisions will be made on altruistic grounds would skew estimates of the times, places, and extent of the impending changes. Many of these noneconomic benefits are discussed in later chapters because they are integral to understanding the evolving energy situation, but they will not alter the inevitable outcome. Awareness of benefits can accelerate or decelerate the transition but only at the margin. The only necessary condition for a transition to solar energy to occur is that those who use or produce energy will act in their own self-interest, a reasonably safe assumption.

The rapidly maturing solar-power industry needs to transform the discussion from one based on environmental doomsday scenarios (which most pro-renewable-energy arguments center on) to one focused on the wealth that can be generated by accelerating the shift to solar energy. Greed trumps fear, which early movers in Germany and Japan are already learning as billions of dollars of global wealth are created through stock market initial public offerings (IPOs) in 2005 alone.[24] The United States, in particular, has a small window of opportunity to become a world leader in these technologies and to reap the resulting rewards, but inaction in this decade may relegate the United States to follower status in the new paradigm.

The Next Silicon Revolution

In the process of replacing an economy founded on fossil fuels with one founded on a renewable, sustainable energy, the world does not have the time or money to try every possible alternative. The disciplines of research necessitate a broad and open mind, but deployment requires a focus on determining and pursuing the best course of action. Facing limited time and money, we must assess where evolving economics will ultimately arrive and focus available efforts on accelerating and therefore benefiting from that inevitable change. Good public policies, research money, and professional talent should be directed to the dispersal of practical, profitable solutions whenever and wherever they are available.

This book analyzes the solar-energy industry and identifies where the opportunities lay as tectonic shifts in energy economics began to affect the landscape now and for decades to come. This analysis clarifies the most likely avenues for early solar adoption along with the accompanying obstacles. By examining the components of the nascent solar economy—including what drives the solar market—individuals, businesses, and governments can commit resources where they will be most effective and profitable.

The driving lesson of this book is to think of solar energy as an industry and economic reality rather than as a philosophical goal, encouraging a new generation of professionals to be involved. Under current reasonable scenarios, the solar industry is expected to grow by 20 to 30 percent each year for the next forty years, which alone should be incentive to attract the world's best and brightest to the challenge.[25] To become fully functional, though, the solar industry needs to develop all the usual institutional underpinnings, including installer networks, training, standardizations, certifications, and relationships with bankers, financiers, and trade groups. Experience in other industries shows that the faster these institutional underpinnings are put in place, the more quickly an industry can develop.

The coming shift toward solar energy mirrors other recent technological shifts that nearly everyone has experienced. Beginning in the 1970s and 1980s, the shift from centralized mainframe computing to distributed microcomputing created dramatic economic benefits to the end user and ushered in the personal computer, the Internet, and broadband information. More recently, similar transformations have occurred in telecommunications as land-line-based networks are supplemented by (or in the case of developing countries, are passed over in favor of) mobile telephony that does not require expensive land-based grid networks to deliver services.

In comparison, at present the dominant technology for making solar cells involves the manufacture of silicon chips that are nearly identical to the computer chips used in the semiconductor and telecommunications industry. The properties of silicon semiconductors, which so greatly altered the world in a few decades by powering the information technology and communication revolution, is set to do the same in the energy sector. The silicon revolution changed industries radically and quickly in the 1980s and 1990s because the new way of doing things was a better way of doing things. Increasingly inexpensive, fast, capacious, and secure information-handling tools were put directly into users' hands. These tools were hard to invent but easy to use: they packed the results of decades of arcane research in basic science into tools that anybody could plug in, turn on, and operate.

The world today stands on the edge of a new silicon revolution that will provide cleaner, safer, more affordable energy directly to users through the mass production of sophisticated devices that require little sophistication to use. The independence conferred by solar energy is one of the intangible, unquantifiable reasons that this revolution is inevitable. Given a choice between otherwise equal options, most people would prefer to be in control of the resources on which their lives and livelihoods depend.

Like the first silicon revolution, the next one will see industries transformed and massive wealth created. Solar millionaires and billionaires will emerge, and markets may even experience a bubble or two of speculative excitement. However, in the end—undoubtedly within our lifetime—we will arrive at a world that is safer, cleaner, and wealthier for industrialized economies and developing ones and in which solar energy will play a dominant role in meeting our collective energy needs.

II

Past to the Present

2

A Brief History of Energy

The future of the global-energy industry can be understood only through examining the industry's history and current configuration is examined as well as the critical moments in history during which energy sources failed. Though seemingly unrelated, events as varied as the establishment of the earliest societies, the fall of Rome, England's early lead in the industrial revolution, and the outcome of World War II were all directly and powerfully influenced by those societies' intimate relationship to energy. Understanding the fundamental role energy plays in our collective well-being provides a basis for exploring the modern industrial world's total dependence on continued access to energy and highlights the precarious nature of the status quo.

Energy: The Root of Life

Long before humans walked the planet, the life that makes up the earth's biological systems relentlessly pursued two interrelated goals—developing effective methods to attract and absorb adequate supplies of energy (in the form of food) and avoiding being eaten as a source of energy by anything else. From simple cellular creatures to large complex mammals, the very nature of life is to repeat the process of energy absorption and conversion for growth, procreation, and self-preservation, and these behaviors have been deeply embedded into the DNA of organisms through millions of years of Darwinian evolution. From the beginning of life in heated ocean vents, ever greater numbers of more complex life forms appeared and pursued these goals with increasing skill and precision—first single-celled organisms, then small multicellular organisms, and eventually plants and animals. As life forms increased in size and mass, they acquired more advanced neural structures and complex behaviors. This added size and awareness enabled and motivated these organisms to seek out and absorb greater and more efficient quantities of energy. At every level of development, however, one basic need remained constant—the requirement to absorb sufficient amounts of basic energy to stay alive, develop, and flourish.

The first law of thermodynamics, also known as the law of energy conservation, is an ever-present constraint in the struggle to access energy sources. This law states that energy—or rather, matter/energy, for the two turn out to be interchangeable—cannot be created or destroyed; it can only be converted from one form to another. For example, burning wood does not create energy but converts energy stored in chemical bonds within the wood into heat and light. Similarly, when animals digest food, their digestive systems break down and convert the chemical energy latent within the food into alternate forms that are in turn used throughout the body to drive various chemical, electrical, and mechanical processes. At every stage, existing energy is transformed into more useable forms, usually with some degree of loss or waste but never changing the total amount of energy. So if energy cannot be created or destroyed, what is the original source of energy with which life can flourish?

All energy that can be effectively harvested and used by living organisms comes in the form of light, heat, or chemical energy. Of these, the primary sources are light, originating exclusively from the sun, and heat, primarily resulting from accumulated absorption of sunlight by the atmosphere of the earth. These primary energies nourish and sustain the planet's creatures, and all fundamental organic processes were derived from them. The remaining forms of chemical energy that are useful to sustain organic life often appear as simple or complex derivatives of light and heat and are formed from other dead organic matter. Consequently, nearly all energy available today—whether in the form of food, fuel, or direct solar energy—originated from the light of the sun. And the techniques that organisms and societies have developed to power themselves rely almost exclusively on a base of stored solar energy in the organic materials of the planet.

As higher-order organisms on earth developed, they generally fell into one of two broad categories—plants (which absorbed their energy directly from the sun through the process of photosynthesis) and animals (which absorbed their energy from eating some combination of plants and other animals).[1] For hundreds of millions of years, this cycle of life continued as plants and animals absorbed sunlight or each other and converted this food to energy. Larger numbers and types of plants and animals adapted to specific local conditions, which allowed for the creation of a wide variety of complex and robust ecosystems both on land and in the sea, with each generation serving as the food and energy source that fed and nourished the next generation. Over hundreds of millions of years, these ecosystems developed complex interrelationships and webs of life built on the soil and organic material of millions of years of ancestry. As ages passed and geologic conditions changed, some of these ecosystems were lost, covered over by sediment from rivers and oceans or from the debris of volcanic eruptions. Some of these remnants of long dead ecosystems and their captured solar energy (the ones with the right combination of geologic features and temperature) were transformed through a process of oxidation and decay over millions of years to become the fossil-fuel deposits that our modern world relies on today.

Aside from those gradual geological shifts, the growth of life was also occasionally interrupted on a global scale by some cataclysmic event that disrupted the balance of energy and limited organisms' ability to continue to collect or concentrate vital energy. One of the best known of these mass extinctions, though not the most devastating, is thought to have occurred during the Cretaceous period around 65 million years ago by what scientists now generally believe was an immense meteor strike in the area of the Yucatan peninsula. This meteor strike threw globe-encircling clouds of dust and sulfur into the atmosphere, effectively blocking out the sun for decades. The resulting reduction in available plant life led to an extinction of many of the animal species on the planet, including the largest, most complex, and highest on the food chain—the dinosaurs. Though this was one particularly devastating event, it is by no means unique in history. By studying the geologic record of fossils, scientists have identified five of these mass extinctions in the last 500 million years, each of which eliminated from a sixth to a half of the existing families of plants and animals in the world at the time.[2] The fossil record shows that a number of lesser reductions in both the quantity and diversity of life on earth have occurred over the last half billion years, and nearly all of them can be attributed to volcanic activity, meteor strikes, or other global geologic events, such as rapid climate change.

Learning to Harness Energy

The dawn of humans around 3 million years ago occurred in a world rich with plants and animals of various shapes and sizes in local ecosystems for which they were particularly adapted. Forests covered perhaps two-thirds of the available land mass, and various ecosystems throughout the world contained soil full of mulched organic matter and minerals.[3] The oceans teemed with plants and animals of various forms. Humans began to visit much of the eastern hemisphere, originating in Africa and migrating to Mesopotamia, Asia, and Europe. Using crude stone tools (and a highly developed cerebral cortex), these early inhabitants were able to gain some basic productivity to manipulate their environments and improve their chances for survival and growth. But not until humans discovered three new technologies to harness and direct energy—fire, domesticated animals, and agriculture—did their impact on the planet begin to accelerate.

With the harnessing of fire around 500,000 years ago, humans became capable of controlling the release of the chemical energy absorbed by and locked into plants, most effectively that from wood.[4] Creating and manipulating this source of energy provided early humans with the huge benefit of concentrated and rapid generation of heat and light for warmth, cooking, and later craft applications, like melting metals and hardening clay. Using available supplies of fuel from dead wood, early humans now had reliable and deployable sources of light and heat. Even though the process of combustion consumed the fuel, the ratio of wood to humans was high, so early humans had a negligible impact on the total amount of available fuel resources. To these early innovators, the chemical energy available in trees represented a seemingly inexhaustible source of energy, and nature brought about its annual renewal at a rate that was well above the rate that the fuel was consumed.

The second new substantial source of energy that humans were able to capture was obtained through the domestication of animals. Omnivorous humans had always hunted and killed animals as a source of energy as food. However, the early human innovators' ability, beginning as early as eighteen thousand years ago, to domesticate and eventually harness into motion animals dramatically increased their capability for energy storage, productivity, and concentration.[5] Domesticated animals provided a reliable and predictable source of food and hides, far more so than hunting and gathering alone. Particularly in times of scarcity or extreme weather conditions, these domesticated stocks mitigated life-threatening risks to these early nomads by functioning as mobile stores of energy, and societies that domesticated these animals achieved substantial insurance and economic benefit from doing so. As Jared Diamond points out in his book *Guns, Germs, and Steel,* the region of Mesopotamia and southwestern Asia was blessed by an abundance of domesticable animals, including dogs, sheep, pigs, goats, and cows, which provided the resources necessary for these early communities to develop the basic structures of civilization.[6]

For reasons highly debated in the anthropological literature, around ten thousand years ago humans began to establish larger and more permanent settlements and farming communities in the valley of the Lower Nile in Egypt and subsequently in the Fertile Crescent of Mesopotamia and other fertile river valleys in China and India. These early civilizations were defined by their decision to forego the nomadic lifestyle of following food and moving with changes in climate, and they began to settle where they could develop more reliable sources of food and energy on fertile riverbanks and in deltas. Through gradual improvements in agriculture and irrigation technology, these early civilizations developed substantial farming capabilities. As permanent settlements grew, an inevitable reduction in nearby wild plants and animals occurred, mainly as a result of overharvesting, which steadily increased these societies' reliance on their agricultural sources of energy. The capture of fertile soil and sunlight would dramatically improve human living conditions and lifespan and set the stage for an explosive growth in human capabilities. Armed with these three tools of energy management, early human civilizations began to absorb ever greater quantities of sunlight, trees, soil, fresh water, and animals to propel their rising demand for energy and feed their growing populations.

Energy Shortfalls Interrupt Growth

These new societies provided many economic benefits to their inhabitants, including specialization of labor, efficiencies in production and distribution, permanent dwellings, and mutual protection against outside threats. However, these groups were not always effective at long-term planning to balance supply and consumption of vital energy and food resources. Though all societies experience boom and bust cycles and attempt to plan for such contingencies, early societies occasionally faced such overwhelming setbacks that complete economic collapse occurred.[7] Not unlike the mass extinctions of prehistoric animals, almost all societal collapses of this type occurred as a result of a depletion in available energy resources, which created a disruption in people's ability to capture vital energy needed to survive.

The first major record of an energy shortfall that induced societal collapse came from the Sumerian civilization of Mesopotamia in the third and fourth millennia BCE.[8] The world's first literate society, and an acknowledged leader in farming, craft, and social organization, this society developed innovative and vast irrigation systems that for generations effectively increased crop yields and fed its growing nonagrarian population. However, Sumerian farmers experienced inexplicable declines in their agricultural yields owing to some unforeseen consequences of their agricultural technology.[9] Increases in the land used for agriculture and in the amount of water diverted and used in agricultural irrigation caused water tables to rise underneath their farms and fields. High water tables interfered with the ability of surface water to permeate downward and increased the amount of surface evaporation on the fields, resulting in a gradual accumulation of trace minerals and salts in the soil. As these salt levels began to rise, crop yields began to drop and did so steadily for over a thousand years from around 3500 BCE. Despite a shift away from less salt-tolerant wheat to more salt-tolerant barley, the salination of the soil grew, and eventually Sumerian society could no longer adequately feed its population.

These energy shortfalls were exacerbated by extensive modifications to the local environment. Trees had been cleared for miles in every direction, which led to increased soil erosion and a decline in agricultural productivity.[10] In the face of reduced availability of food, lack of nearby trees and other vegetation, and limits to the additional availability of arable farmland, Sumerian society began to suffer the predictable consequences of inadequate energy supplies—war over limited resources, widespread illness, starvation, and early death. Ultimately, the economically and politically weakened Sumerian civilization was overrun by the Akkadians in 2370 BCE.[11]

This pattern continues through the rise and fall of Rome. Jeremy Rifkin, in his book *The Hydrogen Economy,* describes this process and notes that "Italy was densely forested at the beginning of Roman rule. By the end of the Roman Imperium, Italy and much of the Mediterranean territories had been stripped of forest cover." He goes on to describe how this also led to severe soil degradation and its detrimental effects on crop yields just as Rome was increasingly reliant on agriculture as a source of energy supplemental to its depleted forests. Eventually, the lack of available energy resources played a significant role in Rome's demise as the institutions of the empire collapsed, paving the way for barbarian invaders from the north to conquer the previously unassailable Roman empire.[12]

China also shares a similar historical pattern. In the early fifteenth century, the vast civil engineering projects of Emperor Zhu Di, including consolidation of the Great Wall, reopening of the Grand Canal, and the launching of the mythic treasure fleets of Admiral Zhung He. These projects, along with a war in Mongolia, led to significant natural-resource depletion, such as the clearing of northern Vietnam's hardwood forests for use in those efforts.[13] The resulting devastation to the local farming communities and inability of Chinese and Vietnamese subjects to adequately feed themselves led to crippling revolts and widespread poverty and starvation. Less than a decade later, the empire's economy had collapsed, the ruling elites had dissolved into a civil war, Zhung He's fleet had been burned or left to rot, and China withdrew into its borders for the next five hundred years.

In the Middle Ages, the pattern repeated itself in Europe.[14] Deforestation increased throughout the continent because nearly everything—cooking, industry, building materials, wagons—required the use of wood for energy. The total reliance of human communities and societies on this single source of industrial energy led to its use and harvest well above the natural rate of replenishment. Eventual depletion in local areas and deforestation around major urban centers led to the decreased ability of agrarian Europeans to reliably feed themselves. Ultimately, this loss of basic energy resources provided severe and unyielding limits to Europe's continued economic growth.

This pattern continues throughout human history in different locations, at different times, and with slightly different circumstances, but the result remains the same. A society taps into sources of concentrated energy—trees, soil, and natural food supplies. Then, in a cycle described in 1798 by English economist Thomas Malthus, that society grows until it exceeds its resource base and collapses. The important corollary to Malthus's argument is that war, waged on any account but in particular because of competing demand for scarce energy sources, accelerates the drain on natural energy resources that are already scarce, dwindling, or depleted. This cycle of resource absorption to the point of collapse is what modern society is at risk of repeating in the decades ahead.

Fossil Fuels Enable Industrialization

Without access to the latent energy of fossil fuels, the current age of industrialization would never have been possible and, in fact, can be defined as industrialization principally by the role fossil fuels played. In seventeenth-century Europe, local depletion of wood fuel and the overreliance on natural and renewing energy resources began to shift the economics of energy away from wood and toward coal. Since medieval times, coal had been used as "the primary fuel for industries such as iron smelting, brewing, glass making, and brick production" and as a fuel for heating and cooking in some major urban centers (such as London) that had easy access to local deposits.[15] Coal's inherent advantage over wood, from the perspective of its application to industrial uses, was that it contained a higher concentration of energy—that is, its more dense structure released more energy pound for pound when compared to wood, thus allowing more energy-intensive work applications to be developed. Coal, however, possessed serious limitations as an energy source. Coal is heavy and had to be labor-intensively mined and transported. This was a difficult, dangerous, and comparatively expensive operation for most of the Middle Ages and rendered coal useful only in those high-value applications, such as metalwork, that could justify the additional cost as compared to wood. Around 1700, innovators began to tap into this new source of concentrated energy to help power a growing number of high-value mechanical inventions and industrial applications. Over time, governments and entrepreneurs discovered that underground coal reserves were both unexpectedly vast and increasingly accessible, and coal slowly began to replace wood as a reliable energy source in industrializing countries. England, in particular, was endowed with a relatively accessible type of sea coal that gave an early boost to that country's industrial transformation.

It was the invention of the steam engine that really enabled industrial applications to multiply. In energy terms, the steam engine is a device that captures the energy from the combustion of a fuel (originally wood and subsequently coal and other forms) and uses that energy to convert water into steam, forcing the pressure of that steam to drive an engine (originally a piston-style reciprocating engine similar to James Watt's design, though turbines are usually used today). This engine is then able to perform work by concentrating the converted energy into a constant, reliable stream to a specific point of focus.[16] Steam-engine energy represented a vast improvement over prior unreliable forms of energy (such as human and animal power), and in many cases, it could be obtained at a much cheaper cost than building an equivalent wind or water mill. The steam engine also provided concentrated and constant energy at a wide variety of locations with less dependence on local conditions since the machine and the fuel could now be brought to where the work needed to be done. One of the first tasks to which the steam engine was applied was the removal of water from coal mines, rendering them more accessible to miners.[17] Armed with a new concentrated energy source and a newly developing set of tools and machines to capture this energy, human civilization took its first tentative steps into industrialization.

Most energy applications from the industrial age generally can be grouped into one of two basic categories—stationary and motive applications. *Stationary applications* are those performed exclusively at one location repeatedly, such as cooking, lighting, and most industrial production applications, including the production of electricity. *Motive applications* are primarily transportation based, moving people or things from one place to another.

Prior to 1800, nearly all transportation was driven by human and animal power on land and by wind at sea. The nineteenth century saw the rise of the steam ship and the steam locomotive as the first large-scale uses of fuel energy for motive applications. By using steam engines to turn paddles or screws, innovators in Europe and America learned to deliver steady, concentrated motive power at sea with revolutionary effect. On land, once track was laid, steam engines mounted on wheels delivered enough energy to move loads of unprecedented size at unprecedented speeds. The combination of these two transportation technologies could consistently deliver more goods more rapidly over longer distances and at lower cost than traditional boats or wagons. Like stationary steam engines, these applications began by using wood as fuel and then shifted to coal to exploit its higher energy density and declining cost relative to wood fuel.

Ultimately cost-effective once deployed, these coal-based transportation technologies were not cheap to build initially. The large initial costs of laying a robust network of railroad tracks or of building ports for ships and the lack of available capital limited the rate at which this infrastructure could be installed. Without boats and railroads, coal's useful application was limited to those places where it could be economically transported. In fact, by 1870, coal still made up only 25 percent of the industrial energy used, despite its many advantages, with wood providing the bulk of the remaining energy needs.[18] If the benefits of fossil fuels were to be made directly available where people lived and businesses conducted commerce, a new delivery method would need to be developed.

The Economic Advantage of Electricity

In the seventeenth and eighteenth centuries, theories of "the electric fluid" (the original scientific concept of electricity) and devices to generate and store small quantities of static electricity were developed and tested. In the early nineteenth century, a burst of industrial inventiveness created the first commercial uses of electricity as Volta invented a prototype chemical storage battery in 1800 and Faraday developed the electric motor and generator (which could convert electric to mechanical energy and back, respectively) in 1821 and 1831.[19] By attaching an electric generator to the powerful mechanical steam engines, inventors were able to produce large quantities of electricity in a steady flow, while motors made it possible to apply that electricity to both motive and stationary tasks. The social impact of this complementary pair of energy-conversion tools—and the resulting electrical communications technologies such as the telegraph—would eventually prove revolutionary.[20]

History, however, is full of fascinating technologies that remained marginal until they could be widely commercialized. Electricity was relegated to this category until Thomas Edison propelled it to success. Edison is remembered primarily as an inventor, but he was also an effective entrepreneur who made inventions useful for end users. His impact, fame, and fortune resulted directly from his focus on developing technologies that were marketable, easily deployed, and standardized. In 1882, he commercialized the transmission of electricity by building the Pearl Street Station in lower Manhattan to deliver electric power for lighting offices, and by the end of the year he was supplying power to over five hundred customers using some ten thousand lights.[21] Prior to Edison's commercialization of electricity, lighting had been primarily provided by candles and lamps that burned animal oil or kerosene—all clumsy, dirty, and dangerous. Edison's low-cost and cleaner electric lighting rolled back the night, increasing productivity and safety for the homes, businesses, and entire communities that were served by them.

Edison laid the foundation that others built on to perfect the industry of electricity transmission. Edison chose to commercialize a technology called *direct-current (DC) generation,* against the advice of a young Czech inventor in his employ named Nikola Tesla. Tesla saw that DC transmission was too restricted in distance and voltage for use in large-scale electric delivery and developed *alternating-current (AC) transmission* to overcome these limitations. George Westinghouse, the head of the Westinghouse Electric Company, recognized the potential of Tesla's invention that Edison ignored. Westinghouse began in 1886 to implement the new polyphase AC-generation method in his central stations, using transformers to step voltages up for long-distance transmission and down for local use.[22] This allowed efficient power delivery over long distances and at various voltages and ultimately proved more commercially successful than Edison's DC system. In fact, AC is still the method universally used for large-scale electricity transmission over power grids today. The economic impact of long-distance transmission of electricity is difficult to overestimate. By substituting "coal by wire" (as the grid-distributed electricity came to be known) in place of the labor- and capital-intensive process of hauling coal from mines in mountainous areas to dense urban centers represented a tremendous cost savings and efficiency, as stringing and maintaining electric wires and cables was substantially cheaper than laying railroad tracks. This new economic model of energy—electricity over wire instead of fuel over railroad tracks—meant that the benefits of coal could rapidly be extended to more places and more people than ever before.

As the electricity industry began to grow, the rush to capture these economic benefits by the new electrical entrepreneurs in the cities of America and the rest of the industrializing world caused much duplication of effort as competing companies strung power lines. In the American Midwest, Samuel Insull, the president of the Chicago Edison Company, was one of the pioneering capitalists who realized that the high-fixed-cost nature of the electricity business meant that substantial profits could be generated if the markets were controlled and consolidated.[23] His and others' aggressive efforts to consolidate local and competing private operators into regional monopolies led to the powerful and concentrated ownership of electricity generation and transmission in America, and by 1932 more than 67 percent of the electric generation in the United States was controlled by eight surviving holding companies.[24] Despite the obvious efficiencies of a single-transmission grid infrastructure, the operators would use their monopoly power to charge exorbitant or unfair prices to customers and led the government to regulate the industry as a public utility with the passage of the Public Utility Holding Company Act (PUHCA) of 1935.[25] With this act, the age of utility regulation had begun, the effects and institutions of which can still be seen in the network of publicly owned utilities and regulatory bodies that exist today.

Over the last seventy years, disparate electric grids have become standardized and consolidated, and the physical integration of electricity-transmission systems has increased. Through operating experience as well as occasional spectacular failures in local generation and transmission, power utilities discovered that reliability, maintenance scheduling, and economies of scale could be improved by tying power systems to each other, integrating and sharing resources across local providers.[26] And as it was true from town to town, it was also true at the county, state, and regional levels. This has led to a steady shift from isolated generating plants to regional systems and culminated in the five major intertied power grids that cover nearly all of North America today.[27] Similar growth and integration occurred in the other industrialized countries during the twentieth century, though many of the industrial economies of Europe and Asia experienced significant disruptions and destruction of their power infrastructure as a result of World War II and later had to rebuild them. Today, though, large interconnected electricity grid infrastructures are the norm for all industrial societies.

The Rise of Oil

Along with the rapid growth in the electricity grid system, another important energy infrastructure also developed during the twentieth century. Motive energy—also known as *transportation*—applications started with the advent of the steamboats and railroads discussed previously, but it took the creation of the internal combustion engine to make these applications widely and regularly used. Early automobile designs used a wide variety of engines including those powered by steam, but in 1885 two German engineers, Karl Benz and Gottlieb Daimler, put the first internal combustion engine on a wheeled carriage and started an industry that pervades every aspect of the modern economy. Though Benz and Daimler succeeded in creating the initial technology, another entrepreneur in America, Henry Ford, developed the assembly-line manufacturing practice necessary to bring the prices down to a level that would make these new machines accessible to the general public. One particularly troubling problem in all of the initial designs, however, was how to fuel these new machines. After evaluating a number of options, including kerosene and steam, the fuel of choice became a by-product of the oil refining process called *gasoline*. Gasoline had previously been generated during the manufacturing of kerosene for lighting purposes, an industry that by around 1900 was on the decline due to the rapid growth of Edison and Westinghouse's local power grids and electric lighting.[28] With the growing popularity of electricity, gasoline was usually available cheaply and sometimes was flushed into the rivers when demand was inadequate. Gasoline's use in automobiles created new growth markets for the oil companies beginning in the early 1900s. And as the market for cars grew, the need for new sources of oil skyrocketed, creating a "black gold" rush to search for additional sources of petroleum.[29] A timely discovery of oil reserves in Texas led to a glut of availability in the United States, making operation of the new fleet of cars cheaper and more widely accessible. With affordable mass-produced cars and ample fuel, demand accelerated, and U.S. annual production quadrupled to 3.7 million vehicles per year over the decade from 1915 to 1925.[30]

Oil as a source of energy has certain economic advantages over coal, particularly for transportation applications, because it can be easily transported in liquid form through pipelines and tanks and can be pumped instead of shoveled. A number of interesting anecdotes concerning the economic advantages of oil can be found in Daniel Yergin's book on the history of the oil industry, *The Prize: The Epic Quest for Oil, Money, and Power*. He discusses the strategic impact oil played in the history of the twentieth century both in the modern industrial economy and in the military. For instance, in the early 1900s, a young Winston Churchill, while acting as the First Lord of the Admiralty, pushed to have the entire British navy converted from coal to oil, even though the United Kingdom possessed an abundance of coal and would need to import almost all of its oil from overseas sources.[31] Despite the potential supply risks this caused, Churchill and his supporters recognized that the added efficiency of handling oil-based fuels reduced the manpower required for fueling and operation, and oil's higher energy density in creating thrust led to a British navy that was superior to, and ultimately able to contain, the coal-fired navy of Adolf Hitler.

Oil's advantages as a fuel for military and transportation uses were increasingly obvious to all of the industrial powers, and access to oil drove many of the tactical and strategic decisions on both sides of World War II. The German push into the oil fields of the Soviet Union's Caucasus region near Stalingrad represented an ultimately unsuccessful attempt to capture Russia's rich reserves for the oil-poor German war machine. Yergin also suggests that one of the Japanese military's primary goals in attacking Pearl Harbor was to ensure adequate oil access by crippling America's ability to block oil shipments from the Dutch East Indies to Japan.[32] Ultimately, the failure of the military campaigns of

Japan and Germany, as their militaries ground to a halt with empty fuel tanks, can be attributed partially to a critical lack of fuel.[33]

Oil, however, has had a few glaring drawbacks as the predominant source of energy for the world's critical transportation applications. The largest of these is that oil is unevenly distributed among the countries of the world, with most reserves lying beneath the countries of the Middle East and specifically not in the industrial countries that rely on them most heavily. The power vacuums caused by the two world wars of the twentieth century created the independent states of the Middle East, including Iraq, Iran, Saudi Arabia, and Kuwait, and it is under these states that the vast majority of the world's remaining oil reserves lie. Following World War II, several of these countries began to flex their political and economic muscle and at a meeting in Baghdad in 1960 established the Organization of Petroleum Exporting Countries (OPEC). This group, including four Middle Eastern countries and Venezuela (later expanded to eleven countries), was formed with the stated objective to help stabilize world oil prices and create orderly markets.[34] However, supply constraints imposed by the Arab members of OPEC in the early 1970s caused massive price shocks in the West, creating significant economic disruption and providing a graphic example of the risks to industrialized economies posed by actions of such a supply cartel. Today, OPEC still controls about 40 percent of the world's oil production and about 66 percent of its reserves, which leads to the real and potential problems discussed in subsequent chapters.[35]

Natural Gas Fills the Gap

From the industrial countries' perspective, oil-price shocks brought the issue of secure access to vital energy supplies into sharp relief and led to policies that would effectively disassociate fossil fuels into different end uses. In 1978, the United States Congress passed the Public Utility Regulatory Policy Act as a response to oil-price spikes in an effort to reduce American dependence on imported oil.[36] The objective was to reduce the use of oil in electricity generation by limiting the type of fuel new power-plants could use, and the resulting changes fundamentally restructured the modern energy industry. Today, the transportation industry remains almost exclusively dependent on oil for fuel while the electricity grid has moved away from oil in favor of both traditional coal and the growing use of another fossil fuel, natural gas. While some areas of the country such as the Atlantic seaboard still use older oil-fired electricity plants, many areas of the Midwest and West have changed to natural-gas-fired power plants because of their closer proximity to local, domestic sources of that fuel.[37]

Natural gas is a fossil fuel that is composed primarily of methane gas, is formed similarly to oil and coal, and often occurs in the same geologic pockets and reservoirs as oil deposits. Though famously used by Robert Bunsen in the invention of his Bunsen burner in 1885, natural gas was difficult to capture, store, and transport in large-scale applications until the second half of the twentieth century. For many years, natural gas was vented or burned off by miners and oil extractors as a nuisance, but the technology for compressing it and moving it through pipelines began to become economically viable after World War II with improvements in welding, pipe rolling, and metallurgical processes.[38] In the 1950s and 1960s, the United States developed vast networks of gas pipelines reaching millions of homes and businesses. Since then, improvements in extraction, transportation, and cleanliness compared to other fossil fuels have driven natural gas to be a growing important part of the global energy mix.[39] Natural gas is perhaps the most versatile form of fossil fuels, and is used in heating applications, as a feedstock for everything from nitrogen fertilizer to methanol to plastic, and as a direct fuel in many new "clean-vehicle" programs.

Shifting Dominance in Energy

The various trends discussed above have driven changing economics of fossil fuel over time and led to a changing mix of sources providing energy to our modern world. Figure 2.1 shows the relative contribution to primary energy of each of the major sources, including traditional renewables (primarily wood), coal, oil, and natural gas, and how they have changed over the last 150 years.

As the figure shows, wood's historic dominance eventually gave way to coal as a primary source of energy, particularly as coal began to be harnessed in both motive and stationary applications. The dominance of coal in the first half of the twentieth century was subsequently eclipsed

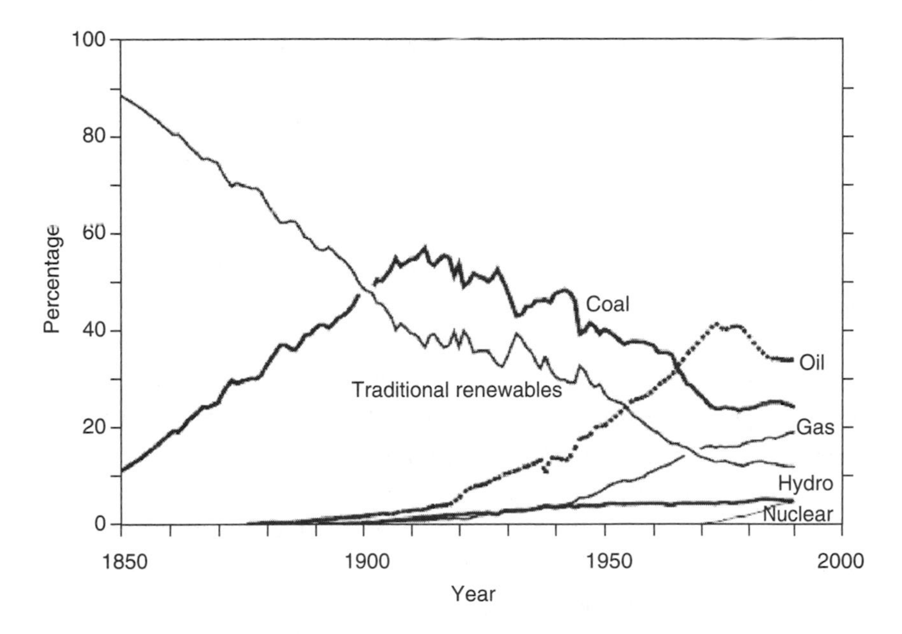

Figure 2.1
Global primary energy substitution among various sources, 1850 to 2000.
Source: Nakicenovic and Grubler (2000).

by oil, which was eventually supplemented by natural gas and nuclear power. What the changing contributions of primary energy sources over the last 150 years show is that rising total energy demand and changing relative economics of these fuels lead to substitutions among them and a pattern of constantly changing dominance in global energy supply. As demand for energy continues to grow in the decades ahead and the relative economic characteristics of various sources of energy change for many reasons (discussed in more detail in subsequent chapters), it is reasonable to expect that the pattern of changing dominance among various energy sources will continue.

An Overview of Modern Energy

Today, there are at least three separate major energy infrastructures on which modern industrial economies rely—the electricity grid, oil refining and distribution, and natural-gas pipelines. Each of these serves a slightly different type of end use for energy, roughly corresponding to electricity, transportation, and heating applications, respectively. Economically, however, maintaining these three separate energy infrastructures requires large annual investments of capital and corresponding large organizations to afford and manage them. Today, six of the twelve largest companies in the world are fossil-fuel providers, with four of the remainder producing automobiles and trucks and one, General Electric, heavily involved in making power systems and wind turbines.[40] Collectively, these eleven companies alone have over $2 trillion in annual revenues, equal to about 4 percent of global gross domestic product (GDP).[41] Total energy expenditure comprises somewhere between 7 and 10 percent of global GDP, and between $4 and 5 trillion is spent worldwide each year on modern forms of energy and electricity.[42]

Global energy production has grown by around 2 percent annually over the last thirty years due to an increase in global population and offset by a decline in per capita energy use or the amount of energy used by each person.[43] Growth has been generally consistent over this period, with only a small hesitation around the time of the oil-price shocks in the 1970s when energy-efficiency initiatives had some limited success and the price of fuel caused people and businesses to temporarily alter their consumption and retool their energy generators for new types of fuel.

Figure 2.2 shows the relative breakdown of the fuel sources consumed in the United States and the sectors in which they are used. The three fossil fuels together (coal, oil, and natural gas) provide about 86 percent of U.S. industrial energy produced, with coal and natural gas dominating the stationary applications and oil dominating transportation.[44] While other nations and regions use different proportions of these fuels for energy, global economic growth and societal well-being currently remain completely reliant on the reserves of latent solar energy in these fossil fuels, with 80 percent of global energy supplied by fossil fuels.[45]

Geographically, total energy consumption is spread unevenly from country to country. The United States is the largest energy user at 26 percent of total consumption, despite having only 4.6 percent of the world's population, and on a per capita basis is one of the highest of any country in the world.[46] For comparison, U.S. per capita energy consumption is more than twice that of western Europe. Collectively, the thirty indus-

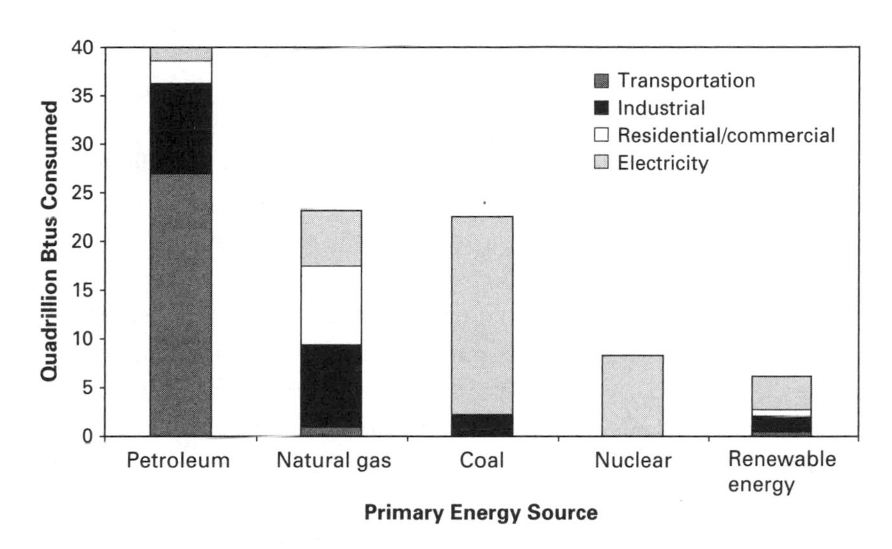

Figure 2.2
U.S. primary energy consumption by source and sector, 2004 (quadrillion Btus). *Source:* EIA (2004).

trialized members of the Organization for Economic Cooperation and Development (OECD), consume over half of global energy, though they contain less than 20 percent of the global population.[47] To visualize the energy gap between rich and poor nations, the average U.S. citizen consumes over eight times the energy of a person in sub-Saharan Africa, even when traditional fuels such as wood and manure, much used in Africa, are included.[48] And some 1.6 billion people, over a quarter of the world's population, have no access to modern forms of energy or electricity at all.[49]

Electricity is a small but quickly growing component of final consumption. From 1973 to 2003, the amount of fuel delivered through electricity generation grew by over 170 percent, significantly faster than the growth in basic energy demand.[50] The growth in electricity use over this period was driven by the rapid industrialization of the modern economy bolstered by the disproportionate growth in wealth and manufacturing in many Asian countries. Looking at these numbers, electricity's importance to the energy infrastructure is understated. In the United States, for example, even though electricity comprises only 18 percent of the final consumption, it requires some 39 percent of the primary fuel supplied—losing some 65 percent of the energy content of its fuel during generation and transmission.[51] As of 2004, the world possessed approximately 3,900 gigawatts (GW) of peak electricity generation capacity, which is used to provide some 16,600 terawatt hours, or 16 trillion kilowatt hours, of electricity every year.[52] (For a description of the metrics of peak capacity and electricity generated, see the appendix.)

Despite the rapid growth in electricity use, the mix of fuels used to generate it has shifted significantly in the last thirty years. Figure 2.3 shows the relative contribution that the various sources of primary energy contributed to electricity generation in 1973 and 2003. Primarily as a result of industrial economies' changing energy priorities after the oil shocks in the 1970s, oil's share in electricity dropped from 25 percent in 1973 to around 7 percent in 2002. To compensate, natural gas grew from 12 to 20 percent, and nuclear power grew from 3.4 to 16.6 percent of electricity generation over that same period. Coal has remained a dominant source of electricity, supplying nearly 40 percent today, and the three fossil fuels combined provide nearly two-thirds of electricity generation.

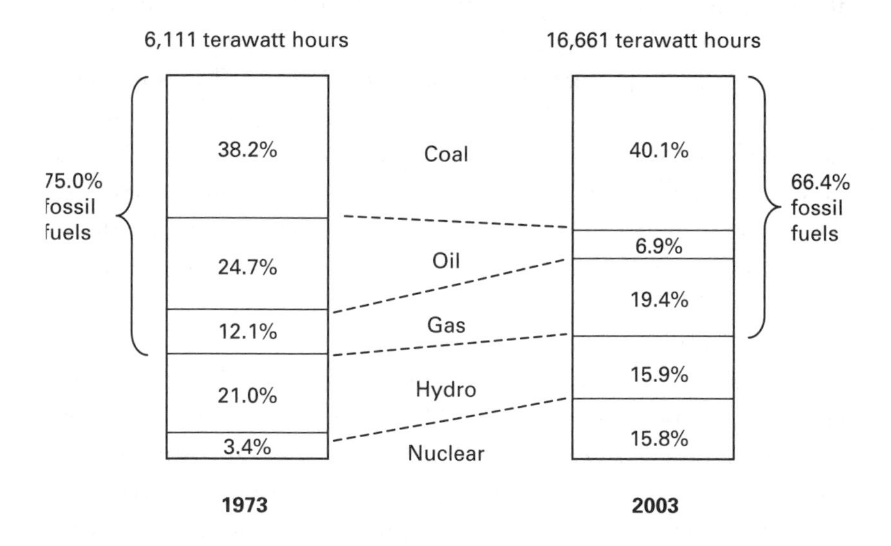

Figure 2.3
Global fuel shares used in electricity generation, 1973 and 2003.
Source: IEA (2005).

This, then, is the current state of energy. Harnessing ever greater quantities of fossil fuels over the last three centuries for industry, transportation, and electricity has allowed for unprecedented growth in the world economy, extended life, and improved livelihoods for billions of people, though not equally around the globe. Owing to the inescapable nature of energy in the modern industry and transportation and the dominance that fossil fuels play in the current energy configuration, industrial economies have to be vigilant about access to and cost of vital supplies of fossil fuels. Chapter 3 examines just how large a risk these factors pose.

3

An Unsustainable Status Quo

From the beginning of the industrial revolution three hundred years ago, electricity, powered transportation, and the use of industrial energy have created an unprecedented improvement in average living standards and human well-being around the world. Increasing efficiencies in the extraction, storage, concentration, and transformation of the planet's natural-energy stores have allowed rapid improvements in both the length and the quality of human life, along with a resulting global population explosion. The combined effect of these two forces—increasing use of energy and growth in population—has also negatively affected the planet and its ecosystems. Natural-resource depletion, social disruption, and environmental degradation are perhaps not surprising effects given the historical precedents of earlier civilizations discussed in the previous chapter.

Awareness of the environmental consequences of the current mix of global energy sources is growing, and unlike similar human-induced environmental changes throughout history, air pollution, climate change, and loss of farms and forests are now global rather than local in scope. In fact, the world is dangerously close to exhausting its capacity for renewal. Until alternative energy sources are developed and deployed on a global scale, society's current pattern of energy use will accelerate the strain on the natural-resource base and on the entire environment, with foreseeable repercussions on economic growth and social stability. Few resources or ecosystems have not been dramatically modified or have not come under threat. The highlights of a recent World Development Report from the World Bank are not encouraging: "Air: polluted, Fresh water: increasingly scarce, Soil: being degraded, Forests: being destroyed, Biodiversity: disappearing, Fisheries: declining."[1] Projecting these trends over the next fifty years and an additional 3 to 4 billion global inhabitants shows that problems we face require serious and immediate solutions if the complete destruction of our global resource base is to be avoided.[2]

In part because of these environmental factors, modern industrial economies' reliance on fossil fuels to provide vital energy has created a situation of substantial economic risk. Growing energy demand and peaking fossil-fuel supply in some regions have the potential to cause a rapid increase in energy prices, not only for oil and natural gas, but for coal as well. Access to the remaining sources of fossil fuels will be threatened as the concentration of remaining reserves in unstable regions grows, creating the potential for both higher energy prices and higher volatility in those prices. Strained and aging energy and electricity infrastructures in industrialized countries add to the likelihood of severe economic impacts. These economic drivers will increasingly compel the development and deployment of alternative sources of energy.

More People Demand More Energy

Today the world is inhabited by some 6.2 billion people, up from less than 1 billion people at the beginning of the industrial revolution and a billion more than fourteen years ago.[3] This rapid growth in global population over the last three hundred years has resulted from two primary factors—decreasing infant mortality and an increasing average lifespan owing to improved medicine and nutrition. As a result of these trends, the industrial period has experienced rapid population growth, peaking in the late 1960s at a global growth rate of 2 percent per annum before dropping to a 1.2 percent increase per annum today.

Population growth is not distributed evenly around the world. Since the 1960s, population growth in many of the wealthiest nations of the world, specifically the thirty member states of the Organization for Economical Cooperation and Development (OECD), has dropped to or below the replacement rate. Not counting the net effects of immigration on these countries, some (such as the United Kingdom, Japan, and Italy) have seen population growth halted or even reversed.[4] Consequently, the growth in global population today occurs almost exclusively in the poorest nations of the world. Despite the $10 billion spent annually on family planning and contraception programs, fertility rates in the developing world remain stubbornly above the replacement rate.[5] Even so, stabilizing the fertility rate is not enough to halt population growth immediately because of an effect known as *population momentum.* Population momentum occurs as the total population lags a couple of generations behind the replacement rate until the proportion of young and old people comes into equilibrium, and this effect limits the effort to slow global population growth. As international family-planning programs continue to slow down the global fertility rate, the momentum effect will drive most of the projected population growth over the next fifty years.[6]

Barring devastating disease or disasters, the global population will grow by nearly 4 billion through 2050, with a substantial majority of these people being added to the 5 billion already living in the less developed regions of the world. From an energy-resource perspective, with 1.6 billion people located in these regions already lacking adequate access to energy, providing enough additional resources to maintain or expand their quality of life will be daunting.[7] Even after spending almost $40 billion annually through the 1980s and 1990s, global rural-electrification programs have barely kept pace with population growth.[8] Expected growth in energy demand from population growth alone will strain the capacity of local resources to keep up and limit the ability of these regions to grow economically. Already, developing nations' ecosystems—including forests, soil, and water—are being rapidly depleted. Continued degradation of these resources, along with the projected population growth, will dramatically increase human suffering unless alternate sources of energy for the people in these areas are developed and deployed.

Although people in the industrial world may try to convince themselves that these issues represent developing-world problems with few local repercussions, the linkages between poverty, war, and the environment prove otherwise. In the post-cold-war era, resource scarcity and security as well as wealth gaps between rich and poor have been leading causes of terrorism and war, both within and among nations. Michael Klare, in *Resource Wars: The New Landscape of Global Conflict,* identifies nineteen ongoing territorial conflicts as of 2002 regarding oil reserves alone, not including the war in Iraq, and many more if conflicts over water rights are included.[9] Every war in a developing country creates direct and indirect costs that also affect the wider international community. For example, the performance and productivity of global industry depends on raw materials and labor from the developing world. During times of war, protecting access to these materials and labor pools or switching to alternatives creates an expense that must ultimately be paid in either inflation or lost productivity. Quantifying these costs in the aggregate is nearly impossible, but trillions of dollars annually would not be an overestimate of the direct military and foreign aid, indirect resource depletion, and loss of economic productivity involved. In a globally integrated economy, there are no purely local effects.

The Environmental Consequences

Increasing global population, energy usage, and energy demand over the last hundred years have already led to environmental degradation of the planet on an unprecedented scale. The global environment today is under siege by human society's patterns of energy absorption, which cause effects such as greenhouse-gas emissions, oil spills, toxic and nontoxic waste, air pollution, inadequate food and water for particular human populations, and loss of animal habitats and biodiversity. Much of the real cost of the damage done today will be borne in ways that cannot be measured, such as lost economic opportunities, increased human suffering, or a world irrevocably stripped of its diversity. The burdens endured by families devastated by increased storm activity, drought, or heat waves or by military activity in unstable regions to protect dwindling energy supplies are never directly linked with the energy policies and practices of prior generations. However, that is often where the root causes lay.

Air Pollution

Various by-products are created in the process of burning fossil fuels to release energy, some resulting from impurities in the raw fuels and others from the combustion process itself. Carbon monoxide, carbon dioxide, and surface (or *tropospheric*) ozone (not to be confused with the *stratospheric* ozone that shields us from the sun's ultraviolet radiation and is not a pollutant) are three of the largest emissions by volume from the fossil-fuel combustion process. Locally, the effects that these pollutants are having on our environment include American cities that are subject to summer smog alerts that require people to stay indoors or refrain from outdoor physical activity to safeguard their respiratory health.

Local air pollution has existed for centuries in urban centers, originally as a consequence of burning fuels for cooking and heating. As far back as 1272, Edward I of England tried to ban the burning of coal in London to reduce the cloud of irritating black smoke that permeated the city.[10] Not surprisingly, this attempt to limit the use of a primary energy resource failed in the absence of a cost-effective alternative. Unfortunately, wood fuel was too expensive for most people to afford even in the face of cold and starvation, so the king's threat of torture and execution had little effect on curbing coal use. Continued coal use has caused London's air quality to deteriorate for centuries. The term *smog* was coined in London around the turn of the twentieth century to describe the mixture of coal smoke and fog, and high-smog days became known as "pea-soupers." A period of four particularly smoggy days in 1952 that killed around twelve thousand residents became the catalyst for serious air-pollution reform in London.[11]

Unlike London's domestic cooking and heating fires, Los Angeles' geography and reliance on automobiles were the causes of its smog. Prior to 1943, smog did not exist in Los Angeles, but a postwar housing and construction boom and migration to the southern Californian desert created rapid urban growth in both population and car ownership. Ever increasing emissions from these automobiles collected in the basin of the San Fernando Valley, were unable to quickly dissipate, and eventually created the worst air pollution in America. Beginning in 1966, California limited tail-pipe emissions in an attempt to address air quality issues, followed by lead limits in 1976. As California added strict auto-manufacturing guidelines to these other antipollution initiatives, Los Angeles reduced the number of days with air-pollution alerts from 120 in 1970 to zero in 1999.[12] Unfortunately, some of these gains in Los Angeles have been reversed in the last few years, as a shift in ownership toward larger and less fuel-efficient vehicles has led to declines in average vehicle fuel efficiency and resulting air quality.

Despite some local successes, the number of people worldwide living with air pollution is increasing, and the human toll of this local air pollution is high.[13] In Great Britain, the National Society for Clean Air estimates that twenty thousand lives a year are shortened by air pollution. In many cities in China, where smog is exacerbated by coal-generated electricity, lax vehicle emissions standards, and increased desertification of farmland, the sun is effectively blotted out for weeks at a time during summer. In 1997, the World Bank estimated that, if air quality did not improve, Beijing would suffer eighty thousand air pollution deaths by 2020, and the city of Chongqing another seventy thousand.[14] The World Health Organization and the World Resources Institute have estimated that over 700,000 deaths worldwide are caused by air pollution each year and that this number could rise to 8 million per year by 2020, primarily in the poorest countries of the world.

Substandard urban air quality no longer affects only urban centers or even the nations that create it. In 2002, the United Nations Environment Program (UNEP) reported the discovery of what it termed "the Asian brown cloud,"[15]—a two-mile-thick cloud of air pollution that covers much of south Asia and the Indian Ocean and is created when fuels are burned for cooking fires and industry. The cloud is changing the local climate in dramatic ways and has the disturbing consequence of both increasing surface air temperatures and decreasing the amount of sunlight that can reach the ground. Other potentially disastrous consequences of the cloud include altered rainfall patterns and reduced freshwater availability, reducing traditional crop yields in a region where food production is already strained.

Food Availability, Desertification, and Deforestation

Since World War II, striking increases in agricultural productivity have been made through the use of various technologies, including tools, seeds, and fertilizers. The so-called Green Revolution has enabled worldwide population growth, even as the amount of land and water available for agriculture has become severely strained. Concurrent with shifting climates and changing rainfall patterns, global agricultural productivity has to deal with resource-depletion problems, including loss of soil and crop failures.

Topsoil, like plant and animal species or fossil fuels, is a nonrenewable resource that is essential for the ongoing survival and prosperity of human beings. The world is losing at least 50,000 square kilometers of arable farmland and rangeland, or about 0.3 percent of the global total, per annum to wind erosion, water erosion, salination, and desertification.[16] This rate is doubling every twenty years or so, particularly as marginal and unsustainable cropland is cultivated in an attempt to maintain agricultural yields. In addition to locations where arable farmland is completely lost, an additional 36 percent of the world's cropland is also experiencing decreased output and productivity from topsoil loss, limiting total agricultural output.[17] Annual grain production per capita worldwide peaked in the mid-1980s and has dropped about 11 percent overall since then.[18] World grain stocks have dropped consistently since the 1960s from over four hundred days of consumption to around fifty-nine days currently.[19]

Focusing on averages misconstrues the extent of the problem, as many locations are at far greater risk than others. In the next thirty years, for example, Africa and Southeast Asia will witness some of the world's worst soil erosion combined with the largest projected population growth. Famine and human suffering will likely increase in these regions as basic access to food continues to dwindle. And with world grain stocks at low levels everywhere, all countries are at risk of harvest failure owing to inclement weather or blight.

Closely related to loss of farmland is deforestation, which contributes to soil erosion, loss of species habitat and biodiversity, and global warming. The pace of deforestation today is astounding. The Rainforest Action Network estimates that in tropical forests alone the rate of destruction is about 2.4 acres per second or 78 million acres (an area roughly the size of Poland) per year.[20] These tropical forests, while comprising only 6 percent of the world's land area, contain 50 percent of the world's species.[21] As forest trees are cut down, habitat is destroyed and the remaining forest's ability to support a diversity of plant and animal species is decreased, and many species die out.

This loss of forest habitats has been a primary driver of what some scientists term the *sixth great extinction*. According to two recent studies funded by Britain's Natural Environment Research Council, the world is now undergoing a mass extinction similar to the five major extinctions of the prehistoric era discussed previously. Plant and animal species are dying out at hundreds of times the normal rate, with every sign of continuing and potentially accelerating. Historically, extinctions of this magnitude were caused by externally imposed events such as massive earthquakes or meteor strikes. Today, extinction is occurring as a result of habitat destruction, pollution, and climate change created by humans and their patterns of energy absorption.[22]

Global Climate Change

Air pollution from fossil-fuel combustion combined with changes in land use are not just changing local and regional conditions but are beginning to damage global atmospheric conditions in ways that will affect nearly all ecosystems and human beings over the next century. Carbon dioxide is foremost among the several greenhouse gases released primarily by fossil-fuel consumption that are radically altering the environment worldwide, including increasing average surface temperatures, changing climate patterns, and sea-level rise. Human activity, including the burning of fossil fuels, has increased the concentration of these greenhouse gases in the earth's atmosphere by 50 percent from preindustrial levels and the amount of carbon dioxide in the atmosphere is higher today than at any time in almost half a million years.[23]

Over the last twenty-five years, the scientific community and the governments of the world have mobilized an unprecedented effort to collect, refine, and interpret data to help determine the role that greenhouse gases are playing in changing global climate. These efforts have been coordinated through the Intergovernmental Panel on Climate Change (IPCC), which was set up by the World Meteorological Association and UNEP in 1988 to evaluate scientific, technical, and socioeconomic information about this threat to the planet and its inhabitants. What these groups have discovered is that global climate change is primarily a result of the global pattern of energy consumption, with up to 85 percent of the emissions that lead to climate change arising from the burning of fossil fuels and most of the rest coming from human changes in land-use patterns.[24] As much as half of these greenhouse gases come from the generation of electricity alone, in large part due to the dominant use of coal, which emits significantly more greenhouse gas than oil or natural gas for an equivalent amount of energy.

Using the best data and statistical techniques available, the IPCC has determined that the burning of fossil fuels, clearing of land and forests, and industrialization have caused a massive shift in the ecology of the planet in the last three hundred years. The IPCC has measured the history of human-induced climate change and documented more hot days, fewer cold days, longer growing seasons, retreating glaciers, melting permafrost, shifting animal habitats toward the polar regions, and so on. Given the momentum of energy use and the continued accumulation of carbon, the IPCC expects that these trends will continue and increase in severity for the next century. It forecasts other destabilizing effects, as well, including more heavy-precipitation events (that is, floods), more violent weather and hurricanes, and increased frequency and severity of drought.

The cost of rapid climate change is most easily measured in the increased frequency of natural disasters. Munich Re, a global reinsurance company, puts out annual reports of natural disasters' cost and severity. The year 2004 was devastating worldwide, with economic losses from natural disasters over $145 billion, up from $60 billion in 2003.[25] Even when the economic impacts of nonweather-related events such as earthquakes and the tsunami in December 2004 are excluded, the weather damage in 2004 caused over $100 billion in economic losses, more than double the year before, owing to increased floods and storm activity. Severe weather events, such as the dramatic hurricane activity of Florida and the U.S. Gulf Coast in 2004 and 2005, respectively, continue to increase, and the number of strong storms (category 3, 4, or 5 hurricanes) in the Atlantic each year has tripled since 1960.[26] According to Munich Re's estimates, the rate of natural disasters has more than doubled since the 1960s, and economic losses have grown more than sevenfold.[27] They further project that by midcentury these types of storm- and weather-related losses will be significantly higher.

Rising temperatures may ultimately be the most costly effect of climate change. A history of global temperature change can best be shown graphically. Figure 3.1 shows the history of carbon emissions, carbon dioxide levels in the atmosphere, and average temperature since the year 1000. The spike in the last one hundred years shows an increase of 0.7 degrees Celsius to a level half a degree higher than any time in the last thousand years. Because of the time lag of climactic effects and the likelihood of increased greenhouse-gas generation for some time to come, the most conservative estimates by the IPCC show climactic effects in excess of and more rapid than those experienced thus far—somewhere

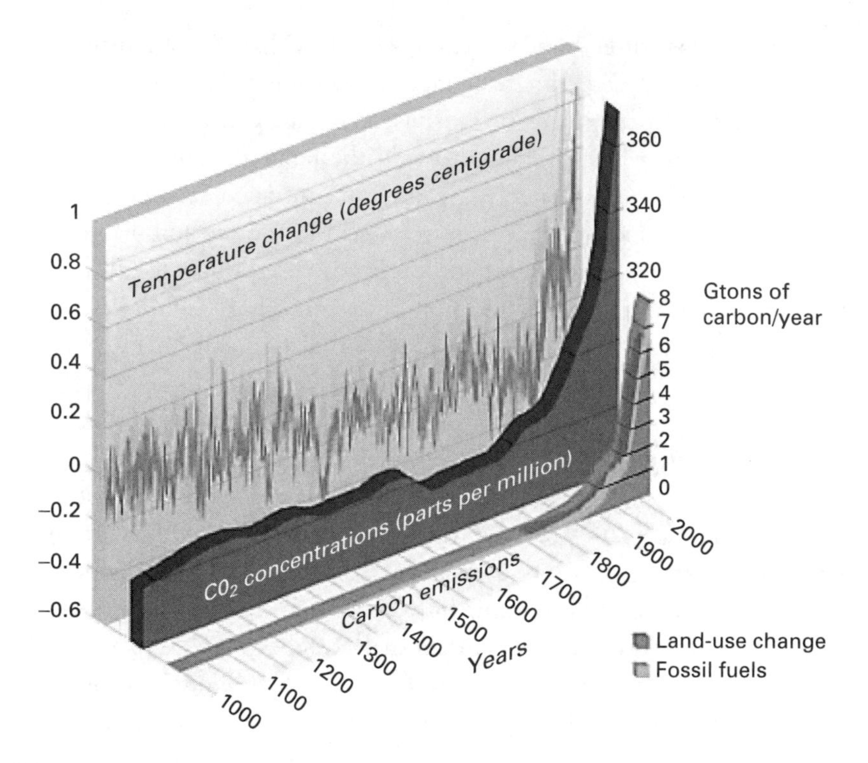

Figure 3.1
One thousand years of correlated changes in global carbon emissions, carbon dioxide concentrations, and temperature.
Source: ACIA (2004).

between one and five degrees Celsius of additional global warming over the next hundred years.[28]

The global-warming component of climate change is troubling both because of its role in the increasing rate of natural disasters (mentioned above) and also because of its direct effects on ocean levels. The IPCC has estimated that the thermal expansion of ocean water and melting glacial ice in Greenland and Antarctica have been the primary reasons the sea level has risen by one to two millimeters per year during the twentieth century and could rise as much as 1 meter over the next century. A one-meter rise would threaten the Maldives, whole sections of the Everglades in Florida, and many coastal cities. According to a study by Geoff Jenkins of the Hadley Center for Climate Prediction and Research, millions of people in island nations will be flooded out.[29]

Potential system-collapse failures are the most troubling of all. There is evidence that environmental regulatory systems such as currents and rainfall patterns are subject to occasional collapse and that the rapidly increased pressure on weather systems from global warming could trigger such a failure. For example, the collapse of the Atlantic Gulf Stream, the oceanic current that brings warmth from the tropics to the temperate regions of the Atlantic Ocean, would cause many climatic changes and have devastating effects on the rich agricultural areas of the northern temperate zone, including much of those in Europe, with severe implications for world health and prosperity.

Today, the scientific debate is no longer about whether climate is changing or in which direction or whether human activity is contributing significantly to these changes; it is only about how fast they are happening and what specific changes each place will experience. It is also now clearly understood that the solutions to these potentially devastating impacts must address the source of the problem—continued reliance on burning fossil fuels for energy.

Energy Scarcity and Price Volatility

Setting aside the environmental issues, the modern economy and the world's ability to adequately support 6.2 billion people at their current levels of prosperity depends on industry's continued access to energy, primarily in the form of fossil fuels, at a reasonable cost. Until renewable alternatives are deployed widely enough to make an impact, three questions regarding the continued use of fossil fuels address the risks that our current energy system poses: (1) how long will the global supply of these finite fuels last, (2) what economic effects will occur as available reserves decline, and (3) what obstacles affect access to the remaining reserves of these fuels, and what would be the consequences of potential disruptions?

The Myth of Oil Reserves

One of the most hotly debated topics in the field of energy is the expected life of the world's reserves of fossil fuels. No one disputes that these non-renewable sources of energy are finite, but there is little consensus about when their extraction will peak and how long these reserves will last at projected usage rates. Whether future supply will be adequate, of course, depends on the level of future demand. Forecasts from the U.S. Energy Information Administration (EIA) predict that global energy consumption in 2025 will be 60 percent higher than in 2001 (about a 2 percent annual growth rate), with the rate of increase in developing countries about three times that of the industrialized world.[30] The EIA estimates that during that time period, fossil fuels will still comprise the 87 percent of the total energy supply.[31] These forecasts are examined in more detail in chapter 6 but may turn out to be optimistic when the expected availability of various fossil-fuel reserves is factored in.

Oil is the fossil fuel with the shortest expected life of reserves and therefore the one most likely to experience shortages first. The World Energy Assessment, a publication sponsored by the United Nations and the World Energy Council, predicts that at projected usage rates, conventional oil reserves will last for another forty to sixty years, depending on the pace at which new reserves are discovered or developed. In fact, since the dawn of the oil age at the beginning of the twentieth century, there have been repeated predictions that the world was running short of oil, which have not yet come true.[32] In reality, the world simply will never completely run out of oil because there will always be some oil available using some recovery and processing technology at some cost. The problem with assessing oil adequacy by looking at reserve life is that this type of analysis ignores a more imperative issue. The question is not when oil supplies will be exhausted but what happens when the supply of cheap and easily accessible oil peaks and begins to decline over the first two decades of the twenty-first century. What happens to the global economy when increasing demand for vital fossil fuels meets peaking and then declining supply?

Peaking is one of the most misunderstood aspects of the oil-sufficiency debate. The basic concept of peaking is that any given oil deposit, by its geologic nature, can be extracted at an increasing rate until about half of the deposit is gone. At that point—called the *Hubbert's peak* after American geologist M. King Hubbert, who famously predicted the effect—two things inevitably occur.[33] First, it becomes practically impossible to increase the speed at which oil is extracted from that field, regardless of the amount of capital equipment applied, without damaging the field's lifetime productivity. Water or natural gas can be injected into the ground to force out more oil, but this will raise extraction costs and at best only slows the field's decline in productivity. For example, Saudi Arabia's biggest oil field, Ghawar, which has been producing since 1951, yields 70 percent of Saudi oil output, but despite the injection of 7 million barrels of sea water every day Ghawar's production has is currently declining at about 8 percent per year.[34] Second, the average cost of producing a barrel of oil from the field begins to rise, partly as a function of the more intense processing required for each additional barrel extracted. At least a third of what comes out of Ghawar's wells today, for example, is sea water, and this percentage will rise.[35] At some point, the oil in the field will no longer be economic to recover.

Even more worrisome than local peaking is when whole regions of oil fields begin to peak. In the United States, for instance, national oil production peaked in 1970. Production dropped from 11.3 million barrels per day (Mb/d) in 1970 to 7.7 Mb/d by 2002, despite massive investments to maintain production.[36] The United Kingdom's oil production peaked in 2002. Norway, the world's third-largest producer after Saudi Arabia and Russia, may have peaked in 2004. Mexico may have peaked in 2005. Nigeria, a member of OPEC, is expected to peak in 2007.[37] And Russia, the world's fastest-growing oil producer over the last decade and closing in on Saudi Arabia as the world's largest, will peak around 2010.[38] Worldwide, some 27 percent of pumped oil is from countries that are at or past their peak production, with an additional 15 percent of the world's production from these countries widely expected to move past peak by 2010. Lack of reliable data from the remaining oil producers (including Venezuela, China, and the Middle East) means that accurately forecasting their peak is difficult, although anecdotal evidence suggests that their peaking point is possible within a decade's time.

As production from oil reserves in the North Sea, Norway, and Russia begin to shrink once their production has peaked, global oil supply will become increasingly concentrated in the area with the largest remaining reserves, the Middle East. The result will be an increase in the oil dependency of Western economies on the Middle East, with the OECD increasing its Middle Eastern oil import percentage from 54 percent in 1997 to 70 percent in 2020 if current trends persist.[39] Having the largest reserves only delays the inevitable, however. If current reserve estimates are to be believed, oil production in the Middle East is poised to peak no later than 2025 and perhaps much sooner.

Since peaking will eventually limit the production of all existing oil wells, the only way to delay a supply-demand mismatch is to increase the number of new oil-reserve discoveries. Oil geologists have looked at nearly every likely location for oil deposits with increasingly sophisticated technology, but rates of discovery still decline. Figure 3.2 from Colin Campbell, a leading oil-depletion analyst, shows that the rate of deposit discoveries has been dropping since the late 1960s and stands today at about a third of annual extraction. In other words, for every three barrels of oil being pumped today, only one new barrel is discovered.

This chart is also interesting because the area under the two curves—the cumulative total of all discovery and production—must, at the end of the world's relationship with oil, be equal. With usage continuing to grow and discoveries continuing to fall, therefore, the upward-sloping production trend is not sustainable. The inevitable outcome of the increasing mismatch between oil supply and production will be a dramatic rise in oil prices, which will accelerate the energy industry's shift to alternative energy sources in the next decade—whether we are prepared or not.

Some people still insist that other, more unconventional oil sources can be economically harvested. The oft-touted tar sands of Canada are one of the largest potential sources but are limited by the amount of energy

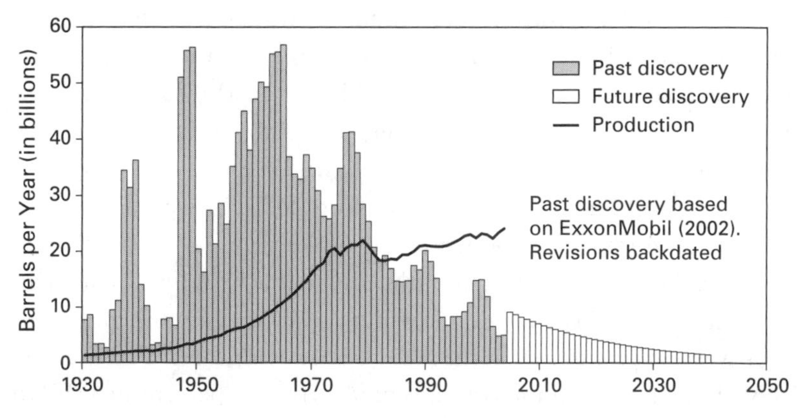

Figure 3.2
The growing gap between annual global oil discoveries and production (billion barrels per year).
Source: ASPO (2005).

required to extract each barrel and the resulting poor economics. These tar sands are petroleum-permeated sands in large surface basins throughout Canada. Petroleum embedded in the tar sands can be extracted through washing and refining, but the process is environmentally dirty and extremely energy-intensive. It requires large amounts of fresh water and natural gas and retrieves only one barrel of oil from two tons of sand.[40] Even ignoring the environmental consequences of this process, as many economic analyses do, the reliance on huge quantities of natural gas will ultimately limit the economic value of this process. At best, these unconventional petroleum reserves may serve as a stop-gap measure to future supply shocks or early peaking of global oil supplies.

Beyond Peak Oil

Since inadequate oil supply is likely to increase its price, could other fossil fuels be substituted to meet global energy needs, particularly in the vital transportation infrastructure? Natural gas, for instance, has been promoted as a clean and widely accessible alternative to oil that would be suitable for in the next phase of our energy economy. The EIA baseline forecast anticipates this increasing reliance when it suggests that usage of natural gas will nearly double from 2001 to 2025 (rising from 23 percent to 25 percent of the total fuel mix over that period), making natural gas the fossil fuel with the highest expected growth rate.[41]

Unfortunately, the value of natural gas as a primary global fuel source is mitigated somewhat by its high cost of transport. Over land, gas can be transported economically via pipeline, but transoceanic delivery has been historically expensive because it is too bulky to be economically transported as a gas. The problem has been partially solved by the development of systems for handling liquefied natural gas (LNG), the product produced when natural gas is cooled to very low temperatures and becomes a liquid, which reduces its volume dramatically and makes it economically feasible to ship by tanker. However, the benefits of LNG are limited by two elements of the transportation process. First, significant energy is required for both cooling and reheating the natural gas into and out of its LNG state, increasing its cost structure and using up to 25 percent of the energy during the conversion process.[42] Second, the capital outlay for liquefying facilities, tankers, ports, and pipelines is prohibitively costly for the amount of energy such an infrastructure can transport, which limits the rate at which this infrastructure can be economically developed. To meet the EIA forecasts, many more such facilities would have to be built to accommodate increased LNG use, but few locations are willing to allow them because of potential safety risks. For example, a 17-million-gallon LNG storage tank—two of which are located within two miles of the center of Boston—packs as much explosive power as three Hiroshima-size nuclear bombs.[43]

The biggest obstacle that natural gas faces, however, is the same peak-supply problem as oil. According to Julian Darley of the Post Carbon Institute, 65 percent of the world's natural-gas producers are already in decline, and worldwide reserve estimates are dangerously overstated, posing a particularly acute problem for the United States.[44] A study by the investment dealer firm FirstEnergy Capital shows that natural-gas production in the Gulf of Mexico, which supplies 23 percent of U.S. natural gas need, declined nearly 20 percent in only two years, from about 14 billion cubic feet per day (Bcf/d) in 2001 to 11.3 Bcf/d in 2003.[45] Even Lee Raymond, chairman of ExxonMobil, has publicly acknowledged that aggregate North American natural gas peaked in 2004.[46] With some 88 percent of new U.S. power plants constructed over the last ten years using natural gas as their primary fuel, peaking in natural-gas production will strain the current economics of electricity generation.[47]

The simple fact is that natural gas is oversubscribed in both number of applications and volumes needed in the coming decades, particularly considering declining reserves and production.[48] Collectively, the supply and demand squeeze on natural gas makes this fuel likely to increase in price over the next two decades. Though it is not a direct substitute for oil in many applications, natural-gas prices have climbed in step with oil prices since the mid-1990s and continue to be correlated.

Coal appears to have a longer future based on known reserves, over two hundred years at current usage rates.[49] However, the amount of coal used for electricity generation and industrial usage is expected to continue to rise from today's levels. At projected rates of growth, reserves of even this vast supply of fossil fuel would be nearly depleted around the middle of this century.[50] Owing to the solid nature of coal fuel, peaking is less of an issue with coal mines than with underground oil reservoirs, but mining coal is expected to become increasingly costly as marginally economic fields are brought on-line to make up for the depletion of larger, more productive fields and as tighter environmental controls on both the mining and burning of coal adds to its net cost.

At the point that the prices of other fossil fuels rise dramatically, substitution with coal will cause devastating consequences on global warming patterns because of its high carbon content and will accelerate the depletion of the world's coal reserves, providing additional upward pressure on energy prices, particularly in countries with large coal deposits (namely, the United States, the European Union, China, and Australia). History has shown that rising overall fuel prices motivate countries to switch energy production to coal, despite environmental costs. Such moves have been seen before: South Africa, in response to antiapartheid embargo pressures, eventually supplied 75 percent of its primary energy production (including 90 percent of its electricity generation) from coal.[51] Faced with similar pressures, more and more nations may opt for using a portion of their coal in liquefaction, or conversion to transportation fuels. (South Africa meets 40 percent of its "oil" needs using liquefaction of coal,[52] and China claims that it will produce 5 million tons of liquid fuels per year from coal by 2008.[53])

In the end, countries faced with fossil fuel shortages, supply disruptions, or high costs for imported energy will look to alternate sources to make up the difference. Economics would suggest that energy users will tend to switch away from expensive productive inputs (fossil fuels) to cheaper ones until the value they deliver become roughly equivalent, meaning that over time the price changes of these substitute inputs will be correlated. Even though each of the three fossil fuels cannot always be easily substituted for each other without substantial conversion and investment, there are enough opportunities to switch among them in applications (such as electricity generation, industrial heating, and transportation) that the prices of all fossil fuels tend to be correlated over time. With all of the major sources of fossil-fuel energy in tight supply, the economic risk of peaking in any one fuel is heightened for all of them.

Supply Disruptions and Price Volatility

Loss of supply through production peaking and decline is not the only way energy prices can be adversely affected. Supply disruptions have the potential to create sharp changes in the price of fossil fuels as well. The risks stem from the disparity between the location of fossil-fuel reserves and where the energy is ultimately needed. Though the industrialized countries of the OECD consume the vast majority of the world's oil, gas, and coal, the sources for this energy are located primarily in the less developed world. While coal is distributed more evenly among industrial economies that rely on its energy (with the United States, the European Union, and Australia together having nearly half of the world's coal), over 93 percent of worldwide oil reserves and over 90 percent of worldwide natural-gas reserves lie under nonindustrialized or non-OECD countries.[54] The economic risk is further concentrated as 78.2 percent of the world's oil reserves are in the eleven OPEC countries and 35.5 and 36.0 percent of the world's natural-gas reserves are in the former Soviet Union and the Middle East, respectively.[55]

Political instability in locations where large oil and natural-gas reserves exist has the potential to dramatically and quickly have an impact on the industrialized world's access to vital energy supplies. Owing to macroeconomic and political events—including war, natural disasters, and intended supply restrictions, such as those OPEC has occasionally imposed—disruptions and price hikes can occur rapidly and unexpectedly. The United States alone has experienced fourteen major disruptions in the foreign oil supply since 1950, including the 1973–74 oil embargo.[56] The cost of these fuels to industrial importers increases as a result of increased fuel-price volatility and the additional cost of protecting access to fuel supplies, either militarily or in the form of foreign aid. Industrialized governments pay tens of billions of dollars per year in the form of military expenditures to protect the security of oil supplies from the Middle East. This estimate does not count the hundreds of billions of dollars spent on the two wars in Iraq since 1990.[57] Though military expenditures rarely show up directly in the pump price of oil or gas, fuel users pay these costs indirectly through taxation and government spending.

As dwindling energy supplies become more concentrated in unstable regions in the coming decades and world demand increases, supply disruptions, for any reason in any geographic location, will have greater impact on global energy availability and prices. Armed insurgencies or regime changes inevitably disrupt supplies until normal levels of production and investment can be reestablished. In Iraq, home of the world's second-largest oil reserves after Saudi Arabia, the armed insurgency against U.S. troops has prevented oil production from recovering from the levels prior to the invasion of 2003. Iraqi oil pipelines and terminals have been bombed scores of times, cutting crude-oil exports from 2.5 million barrels per day under the U.N.-sponsored Oil for Food program to no more than 1.3 million as of late 2004.[58] Although, this loss represented less than 2 percent of global consumption in 2004, the resulting oil-price spikes in the summer of 2004 illustrated the inelasticity of world demand.

Even outside of the Middle East and Russia, the fossil-fuel transportation infrastructure is at risk for disruption. The bottlenecks created by concentrated production and distribution make easy targets, leaving the fossil-fuel industry vulnerable to disruption and terrorist attacks. Attacks on pipelines are common, but refineries and tankers can also be targeted, and the shipping lanes themselves are at risk. The Energy Information Administration identifies seven oil-transit choke points, such as the Panama Canal and the Strait of Hormuz, where a blockage or large-scale attack could cripple whole segments of the oil-distribution system.[59]

Other structural industry factors add to fossil-fuel price instability. First, the world is operating on razor-thin fuel inventories because producers have not reinvested in pumping capacity, tankers, pipelines, and refineries.[60] As of June 2004, OPEC's spare pumping capacity had dropped to 3 to 4 percent of total output; by August 2004, it had fallen to less than 1 percent due to unexpectedly strong growth in demand.[61] The ability to add to oil-production capacity is limited by both local peaking and the long lead times for additional investment. OPEC can no longer be the global swing producer for an unexpected drop in supply or excess demand owing to war or weather, as its members are already producing at nearly full capacity. As a result, even small disruptions or even threat of disruption, at any stage of the supply chain, can cause rapid price spikes.

Risks to the Grid

In addition to the risks of systemic fuel peaking and potential supply disruptions, the infrastructure for turning fossil fuels into electricity and delivering it to end users is also threatened. Electricity grids have always been prone to dramatic failures and potential problems. North America, for example, has experienced the major blackouts of 1965, 1977, 1996, and 2003; various environmental issues in the 1960s and 1970s; the fuel crisis in 1973–74; and the Three Mile Island accident in 1979. There have been few years when the North American energy grid has not had to deal with some crisis.[62]

Beginning in the 1990s, many countries, including the United Kingdom, Chile, Australia, New Zealand, Japan, and Finland, and about half of all U.S. states began to deregulate their electricity-generation and -transmission businesses.[63] The primary problem under the earlier, regulated system was that the utilities that delivered electricity had no incentives to reduce consumer energy usage or their cost structures because revenue was determined on a cost-plus basis. Governments favored deregulation as a way to lower costs through increased competition and a separation (or *unbundling*) of the services provided by electricity providers. In theory, deregulation would drive increased competition and innovation at all levels of the electricity supply chain, but in practice deregulation has developed its own set of poorly structured incentives that still threaten the electricity infrastructure.

Since deregulation, the incentives for power generators and transmission utilities have changed toward short-term profits and maximizing cash flows rather than investing for adequate future capacity. As utilities continue to underinvest in capital and capacity, industrialized economies are subjected to increasing risk of system failure. Currently, the average age of the U.S. electricity grid's components is increasing, with some sections over seventy-five years old.[64] While system failures are part of any electricity system, the size and frequency of these failures are escalating. In August 2003, the eastern United States and Canada experienced the largest blackout in history, affecting some 50 million people. One month later, 57 million people in Italy lost power for a day. Both blackouts were the result of downed power lines and inadequate system safeguards, costing each country billions of dollars in damages and lost business. These recent, dramatic events have heightened awareness that the electricity grids are vulnerable and deteriorating.

Aside from underinvestment in the energy infrastructure, critics argue that poorly designed deregulation has increased the volatility of electricity prices, as exemplified during the California energy crisis of 2000 and 2001. After deregulating only wholesale electricity prices in 1996, the California electricity market began to experience capacity shortfalls owing to unexpectedly strong demand and by 2000 had become tight enough that even small supply disruptions caused a more than tenfold increase in the spot-market price of electricity. As a result, many retail utility electricity providers went bankrupt, and eventually California had to partially reregulate the electricity supply.[65] In Europe, too, waves of electricity outages in the last couple of years have led to widespread discussion about the wisdom and value of deregulation.[66] European public support for deregulation continues to dwindle, and many proposals have been presented in Brussels for new interventions to ensure adequate and reasonably priced electricity supplies for European citizens.[67]

All of these factors—fossil-fuel peaking, potential supply disruption, and an aging electricity grid—put the future of conventional energy at risk in every part of the energy supply chain, from the source to the end user. The world's energy future is increasingly dependent on sources concentrated in the volatile countries of the Middle East and in Russia, and fossil fuels are set to increase dramatically in price as global output of oil and natural gas peak in the next decade. Until the industrial economies are able to make a switch to alternative sources of energy, the transfer of wealth to fossil-fuel providers and damage to world economic productivity will continue. As our current energy consumption further damages the air and atmosphere and leads to unsustainable absorption of land, trees, and fossil fuels, the risks of increased human suffering multiply.

Each of these effects and risks is alarming when considered independently; the prospective combination of them all is nearly overwhelming. There are no obvious areas of opportunity to rely on to address any one or combination of these concerns. No single area gives us some sense of comfort about our ecological or economic future as we deplete our existing resource base. Throughout history we have seen human ingenuity conquer previously insurmountable challenges, which provides some hope that our current concerns will yet be addressed. This hope, however, overlooks the pain and suffering and long time spans many of those transitions have entailed.

If human ingenuity is to provide answers to natural-resource scarcity and the economic consequences of fossil fuels, it will need to find new methods to provide the basic resources on which our lives and prosperity depend—food, water, and energy. While a comprehensive analysis of food and water problems and their solutions is beyond the scope of this book, their causes, effects, and potential responses are intimately linked to adequate future access to energy. Various new sources of energy are being considered. They are explored in detail in the next chapter.

4
The Field of Alternatives

As the previous chapter explored, the growing mismatch of supply and demand of energy and the increasing external costs of current energy practices are providing economic, environmental, and social pressures to develop alternate supplies of vital energy. Many argue that economic pressures can be mitigated by increasing the efficiency of energy use in everything from vehicles to light bulbs to appliances and by reducing energy losses in homes and offices through the more effective and widespread use of insulation in building materials. Increased energy efficiency will, indeed, play a major role in reducing the costs and risks of the global energy infrastructure, and it is often the cheapest and most effective method of addressing such issues in the short run until widespread deployment of alternatives can occur. For instance, the global transportation infrastructure and the existing stock of cars, trucks, ships, and planes (and the petroleum infrastructure necessary to manufacture and fuel them) is so vast that more efficient use of remaining petroleum reserves represents the only viable method of addressing increased energy costs and dwindling supplies in this sector for the next few decades.

In the end, however, efficiency can be only a portion of a new energy solution because efficiency alone cannot solve the problems of both keeping the price of declining energy stocks in check while also providing opportunities for growth in wealth and prosperity for the billions of people beyond the industrialized countries. Simply put, increasing global population, growing industrialization, and declining resource availability are (as they have been since the beginning of the industrial revolution) more powerful pressures than efficiency alone can withstand. Efficiency can and will help to address issues in global energy supply, but vast new sources of energy will eventually have to be deployed to avoid both punitive economic changes and a declining standard of living for fossil-fuel-importing nations. This chapter begins with a review of potential new energy sources that are geared primarily toward generating electricity. (Direct solar technologies are left aside, to be explored in more detail in chapter 5.)

As many of these alternative sources of energy are best suited to generation of electricity, many of the solutions that will eventually arise to meet the energy needs in transportation and heating applications are likely to be direct or indirect applications of electricity. As a result, harnessing sufficient local renewable sources of electricity becomes the necessary first step to address the balance of all global energy issues. The second half of this chapter examines hydrogen fuel-cell technology in terms of both its promise and challenges—how it can be used for both stationary and motive purposes and how it may bridge the gap between electricity and today's fuel-based applications.

Renewable-Energy-Generation Technologies

In reviewing the field of alternative renewable technologies (including nuclear, hydroelectricity, wind, geothermal, and biomass), each has advantages, limitations, and unique economic considerations that determine its value in providing energy for specific applications and locations. Businesses and economies will pursue the alternative renewable technologies that they are best suited for by natural-resource endowment and that they can develop the necessary expertise for fairly quickly and at the lowest cost. This section asks relevant questions about each energy technology: how is it currently being employed, what role does it serve, what type of power does it cost-effectively supply, and what natural limits exist to its growth and deployment?

Each of these technologies needs to be examined not as one class but on its own merits. All of these technologies will have a role in the aggregate energy system of the future because each is better suited than others to meet specific energy needs. Disaggregating these technologies and looking at the costs and benefits of each individually are critical to informed forecasts and decisions about the future of the energy system. Because technological progress and research and development continue to push back technical boundaries and may change the relative economics and viability of each technology, no technology should be dismissed, even when the initial answers are unclear or a given solution appears to be limited.

The most important element for analysis is the expected cost of each technology among its likely applications, and it is no simple matter to make these comparisons. Various methods of production, use, and financing can affect the analysis and are often hard to estimate reliably. It is even more difficult to project many of these variables over the long life of an energy investment, injecting added risks into any such calculation. The true cost of many of these technologies is further obscured by the absence of many social costs in their final energy price. Just as a portion of the cost of securing oil supplies in foreign countries is paid not at the pump but in taxes, many of the hidden costs of nuclear and hydropower are not paid for in the market price of the electricity they deliver. These direct and indirect economic dimensions are examined for each alternative energy technology below.

Hydroelectric Dams

A hydroelectric dam is a striking achievement of human engineering, the sight of which rarely fails to inspire awe. More important, dams are powerful industrial tools. From the United States' Hoover Dam in the 1930s to Egypt's Aswan Dam in the 1960s to the recently completed Three Gorges Dam in China, these massive industrial projects have demonstrated humankind's ability to manipulate the environment to provide useful energy. Traditional hydroelectric dams turn streams and rivers into freshwater reservoirs while generating electricity for local communities and regions. They regulate the seasonal flow of rivers to prevent downstream floods and provide water to local farmers for crop irrigation.

At first glimpse, hydroelectric dams appear to possess compelling economics because they provide large quantities of electricity reasonably cheaply compared to many other forms of energy—between two cents and ten cents per kWh, depending on dam location, size, and type.[1] At the same time, these dams provide locally generated power with no fuel cost and little availability risk. With over 45,000 large dams in operation today, their engineering and economics are reasonably well understood, allowing dam builders and owners in industrialized countries to use capital markets effectively to finance their construction, minimizing lifetime project costs and therefore the electricity they provide. Developing countries, too, have historically been able to finance hydroelectric dam construction through infrastructure loans from international lending institutions such as the World Bank. For these reasons, dams already play a significant role in the modern electricity infrastructure, providing some 13 percent of the electricity generated in OECD countries (19 percent globally), with almost half of the world's rivers containing at least one large hydroelectric dam.[2]

Dams often directly serve the needs of large-scale industrial electricity users such as aluminum smelters and paper mills, which often purposely locate their plants near hydropower projects.[3] Hydropower is also often used to fill the need for *peak power*—the fluctuating 5 to 10 percent of power demand that must come on line quickly to meet rapid and unexpected changes in supply or demand. The inherent advantage of using hydroelectric power for a peak-power application is that the constantly turning turbines provide what are called *spinning reserves* that are ready to generate electricity at any time. These reserves can be quickly activated, taking only moments to come on-line and fill unexpected power gaps, a critical need in the delicately balanced modern electricity grid.

Recent concerns have cooled interest in starting dam construction in most industrialized nations because these hydroelectric projects suffer from hidden costs that make them both economically and environmentally questionable. The first and most obvious of these hidden costs is that dam construction displaces thousands of people and millions of animals as the requisite large reservoirs are filled and former riverside communities and farms are flooded. According to the World Commission on Dams, dams built between 1950 and 1990 displaced some 40 to 80 million people around the world, often without compensation or assistance.[4]

At the same time, the process of filling a dam's reservoir floods substantial areas of farmland, forests, and bogs. Recent research on climate change and global warming has shown that decaying plant matter from these flooded areas contributes to a significant amount of greenhouse-gas emissions. These flooded ecosystems rapidly release the stored carbon in plant material as carbon dioxide and methane gas, potentially as much as generation of an equal amount of electricity through fossil fuels.[5] This research brings into question the historical assumption that hydroelectric power is a relatively clean energy technology. Furthermore, once a dam is built and the flow of water has slowed, silt and soil no longer flow downstream because they are trapped behind the dam. The river then cannot replenish soil that is naturally lost to erosion downstream into the fertile river valleys and deltas used for local agriculture. Dams also damage fish populations, such as the salmon of the American Pacific Northwest, because dams block them from making the necessary journey upstream to spawn in their native grounds. Finally, dams can supply irrigation water to local populations but at the expense of denying water to communities downstream. Many dammed rivers—such as the Colorado River in the United States and Mexico or, for several months a year, the Yellow River in China—no longer flow to the sea.

In many countries of the industrialized world, there are limited additional feasible sites for hydropower, and long lead times for construction mean that they cannot be quickly harnessed to meet changing energy economics. In the United States, for instance, the Department of Energy estimates that if every viable location were developed, the country could increase its hydroelectric power supply by no more than 40 percent—which would equate to less than 5 percent of present U.S. electricity usage.[6] This increase would hardly meet projected growth over even a couple of years, much less indefinitely replace dwindling and insecure fossil-fuel resources.

Growing awareness of the economic and environmental issues that hydroelectric dams face has not only slowed construction of new dams in industrial nations but has also led to a drop in financial lending for hydropower projects in the developing world. World Bank lending for large-scale hydro, for example, has dropped nearly 90 percent in the last decade, from around $1 billion per year in the early 1990s to around $100 million in 2002.[7] While many more viable undeveloped hydropower opportunities exist in the developing world, environmental and political pressures as well as lack of funding render it unlikely that enough of these projects can be completed to significantly increase hydropower's role in worldwide power generation.

Nuclear Power

Nuclear power, which provides nearly one-quarter of OECD electricity needs, plays a larger role in the current worldwide energy mix than hydropower. Thanks to massive investments in nuclear-energy technology beginning in the 1950s, the nuclear industry emerged during that period as the golden child of electricity for industrialized economies. In the 1950s, nuclear energy promised nearly limitless power generated so inexpensively that U.S. Atomic Energy Commission chair Lewis Strauss declared that "it is not too much to expect that our children will enjoy electrical energy too cheap to meter."[8] As a result of this promise and to wean the global electricity supply from oil after the supply shocks of the 1970s, nuclear power grew from 2 percent of world electricity generation in 1971 to 17 percent in 1988.[9] But just as the nuclear industry began to achieve dominance as a major source of electricity generation, three major issues surfaced that limited the world's enthusiasm.

First, increasing awareness of nuclear-energy economics and energy-system dynamics caused a stall in orders in the United States, the world's largest nuclear energy user. In the U.S. market, new plant orders peaked in 1973 and then plummeted to zero from 1979 on, mostly because an excess in electricity-generating capacity made their economics questionable. This oversupply resulted primarily from overestimations by electric utilities of expected electricity demand growth and underestimations of the effects of energy-efficiency programs of the 1970s.[10] Once this oversupply became apparent, utilities stopped ordering new reactors largely because they did not have markets large enough to absorb their existing electricity-generation base.[11]

Second, two headline events heightened public awareness of nuclear power's safety and environmental risks and led to increased safety requirements for new and existing plants. The 1979 near meltdown at the Three Mile Island plant in Middletown, Pennsylvania, and the resulting mass evacuation of the local population showed the potential serious risk from nuclear-reactor failure to local communities. As a result of panic from the Three Mile Island incident, nuclear-power-plant orders ceased in all but three OECD countries, pending further safety evaluations.[12] Then, in 1986, the meltdown at the Soviet Union's Chernobyl reactor spread a cloud of radioactive dust over large parts of the Ukraine and western Europe, particularly devastating the local region nearest the reactor. Indeed, the melt-down at Chernobyl could have been much worse: the prevailing wind blew Chernobyl's plume of radiation away from Kiev, a major Russian city only eighty miles away. The Chernobyl melt-down confirmed that large reactor-containment failure was possible and showed the catastrophic environmental effects that resulted when such a failure occurred.

Third, many nations began to use their domestic nuclear-energy programs as a cover to start or expand military weapons programs. Interest on the part of the international community in seeing nuclear power disseminated globally cooled rapidly when India, which had gotten its start in nuclear-energy technology from the U.S. Atoms for Peace program in the 1960s,[13] set off what it called a "peaceful nuclear explosion" in 1974. Since the 1990s, the threat of proliferation of fissile material from poorly guarded former Soviet warheads and nuclear reactors and the open desire of many Asian and Middle Eastern countries to acquire nuclear weapons have elevated the risk of a second, and much wider, nuclear arms race and made further deployment of nuclear energy in the developing world less desirable.

Only a few countries, such as France (where 75 percent of the domestic electricity comes from nuclear power) and some Asian users such as China, Japan, and Korea are seriously pursuing additional nuclear capacity.[14] In particular, several of the fossil-fuel-poor economies of Asia, such as Japan and Korea, have tapped into nuclear energy as a way to jumpstart their economic development and gain a measure of energy independence. Conversely, no new nuclear plants have been commissioned in the United States since 1978, and six European countries—Belgium, Germany, Italy, the Netherlands, Spain, and Sweden—have committed themselves to phasing out nuclear-power programs already in place.[15]

Currently, the fact that nuclear energy generates electricity without releasing meaningful quantities of greenhouse gases is being cited to justify further plant construction. Nuclear power also has certain long-term economic advantages over fossil fuels—its stable fuel costs and immunity from the supply disruptions that plague imported fossil fuels. The type of electricity that nuclear plants provide is known as *base-load power*—that is, the minimum amount of electricity that must be constantly fed into the grid to ensure uninterrupted electricity operation. The nature of the reactor process, which is expensive to start and stop, both allows and mandates that nuclear-power plants be kept in operation 90 percent or more of the time, thus reducing the cost of nuclear-generated electricity as fixed costs (primarily the cost of building the reactor and facilities) are spread over a larger volume of electricity produced. Nuclear-power plants provide some 24.2 percent of the electricity in OECD nations, despite comprising only 15.1 percent of the installed generation capacity, because they run such a large percentage of the time.[16] Even with this capacity-utilization advantage, however, high construction costs and safety-procedure costs translate into an electricity cost of between six and seven cents per kWh under optimal conditions, which is consistently more expensive than the other fossil-fuel options for base-load power.[17] Other estimates range as high as ten to fourteen cents based on the technology most widely deployed since 1980.[18]

These estimates radically understate the true cost of nuclear-generated electricity by ignoring the substantial hidden costs that are paid for their use. The hidden costs of nuclear power include the cost to dispose of radioactive waste, even assuming that a technologically viable means to do so becomes available. No nation that currently uses nuclear power has developed an adequate solution for the eventual storage of nuclear-waste products. The United States, for instance, is currently grappling with cost overruns and political fallout from its attempt to sequester spent nuclear fuel and waste at the proposed Yucca Mountain repository in Nevada. At the same time, serious reservations about site suitability and the methods of transferring spent fuel to the Nevada site are slowing the approval process. In the meantime, nuclear waste is being held in cooling pools and on concrete pads at the many nuclear facilities around the country, creating significant environmental and security hazards. Further, the eventual decommissioning, disposal, and site rehabilitation of nuclear-power plants impose a hidden cost burden on industry economics that is both difficult to predict and rarely adequately accounted for in estimates of the cost of electricity generated by these plants.

Finally, the cost of electricity generated by nuclear technology ignores the staggering costs of the damage caused and the cleanup and rehabilitation required when the technology fails. It is estimated that cleanup for the Chernobyl melt-down will cost somewhere between $26 billion and $34 billion by the time it is eventually completed, which exceeds the value of all electricity ever generated through nuclear power at all plants in the former Soviet Union.[19] Historically, these costs have never been included in calculations of the cost of nuclear electricity because they are accounted for instead in taxes and government spending. In the United States, the nuclear industry is further shielded from liability for nuclear accidents by the 1957 Price-Anderson Act, which mandates that industry liability for damage from such a failure be presently capped at about $9.5 billion—a small fraction of what a major nuclear accident would cost.[20]

Nuclear power, like large-scale hydropower, cannot really mitigate rapid energy-supply shortfalls because the design, approval, and construction cycle can take as long as a decade. In addition, the global electricity industry is going to face a capacity strain as most of the nuclear plants built in the 1960s and 1970s are decommissioned when they reach the ends of their forty- and fifty-year effective lifespan over the next couple of decades. Attempts are being made to replace aging nuclear plants with new ones or to rehabilitate existing plants for longer lives, but so far these efforts are being fiercely resisted by both the fossil-fuel industry and the mainstream renewable-energy movement. However, in the event that the fossil-fuel infrastructure is jeopardized through dramatic cost increases or supply shocks, nuclear power may be promoted to play a larger role in the energy infrastructure. If decision makers continue to perceive that no other viable option for meeting electricity demand exists, desperation may drive them to accept the risks associated with nuclear energy. For purely economic reasons having little to do with ideology, nuclear energy will ultimately fail to provide a viable solution.

Wind Power

The increasing interest in wind power over the last decade is an example of how quickly an energy technology can grow when it becomes cost competitive. Over the last decade, the amount of wind power installed each year has increased by a factor of eight, a growth rate of roughly 25 percent per annum.[21] Today, the industry represents around $10 billion in annual wind turbine sales.[22] Though still small in the grand scheme of total electricity generation (supplying roughly 1 percent of global electricity), wind's relative market share is destined to increase rapidly. The primary reason for this dramatic growth will be that at favorable sites, industrial-scale windmills have become cost effective compared to all other forms of electricity generation. The American Wind Energy Association estimates that for a large wind farm, electricity can be generated for between three and six cents per kWh.[23] However, large scale is determinative. Large wind turbines produce electricity more cheaply than smaller turbines, and large wind farms are also generally more cost-effective, so the application that is most economic today is an industrial grid-feed wind farm, comprising dozens of very large turbines.[24] Costs for these optimally sized wind farms have halved since 1990 and continue to fall more quickly for wind turbines than for most other renewable-energy technologies, but the rate of improvement is slowing as the technology matures with costs presently decreasing by around 2 to 3 percent per annum.[25]

Europe has taken the lead in the development of wind electricity and accounted for 75 percent of the world's installed wind-turbine capacity as of 2003.[26] The reason, not surprisingly, is economics: Europe's average electricity prices are among the highest in the world because it has high taxes and few native sources of fossil fuels. This high cost structure has motivated European nations to become leaders in developing wind power and has pushed wind to the forefront of the "new renewable" energy technologies. Though wind-turbine technology can be both cost-effective and clean, there are limits on further growth of the industry. The size and location of a windmill or wind farm can cause fierce local community resistance. As discussed, to be cost-effective wind turbines must be large and (often) located in prominent areas, as on the tops of ridges or hills, creating significant visual impairment. This has been a particular problem in U.S. states such as Massachusetts, where zoning and visual-field issues limit wind power's potential market development.[27] The British public has also resisted many new wind-power sites out of fear of visual disfigurement and risks to property values in nearby areas.[28]

To sidestep aesthetic issues and to allow for larger turbines to be deployed, the wind-power companies of Denmark and Sweden have begun in the last few years to construct offshore wind farms, with many more planned in the United Kingdom and elsewhere. According to the European Wind Energy Association, by 2010 Europe will experience a tenfold increase in the amount offshore windpower over 2005.[29] But installing windmills at sea is more capital intensive than on land, and offshore sites have their own set of unique maintenance issues, such as the need for specialized ships to handle large repairs.

Various groups are resisting offshore windpower installations for other reasons as well. For example, Spanish fishermen are fighting a proposed 400-tower wind farm off Cape Trafalgar on the grounds that it would interfere with tuna migration and force small boats to make dangerous detours.[30] A proposed 130-tower wind farm off Cape Cod in Massachusetts has met fierce local resistance on the grounds that its 400-foot windmills (taller than the Statue of Liberty) would kill seabirds and endanger tourism.[31] Despite these challenges, offshore wind farms will likely be a growing part of the renewable-energy landscape in coming years as their economics improve to the level of their land-based counterparts.

Large-scale use of wind to generate electricity is, at the moment, limited by the nature of the wind resource itself. Wind is intermittent, which causes the electricity that wind turbines provide to fluctuate, sometimes dramatically and unpredictably. If the wind speed is too low or too high to be useful or optimal, then a turbine is unusable for electric generation. As a result, wind turbines alone cannot be large-scale providers of electricity to the energy grid regardless of their cost-effectiveness and cannot reliably provide either peak power such as hydroelectric dams or base-load power such as nuclear plants. If they are to be deployed as more than a small fraction of the electricity grid infrastructure, they must be coupled with backup generators or large-scale storage.

Another limitation on wind-power industry growth has been a result of the methods that many governments have used to stimulate wind-technology deployment. Recent years have seen wind-turbine production seesaw in both Europe and America as production tax credits designed to stimulate wind markets have lapsed or been modified. Uncertainty about subsidy renewal has caused sales and shipments in some countries to drop in certain years until production returns full swing as legislative issues are resolved. Even when such a lapse corrects, it has taken time to reprime the production pump for location evaluation, project finance, and permitting. Owing to production tax-credit lapses at the end of 2003 in the United States, companies that sell into the U.S. wind market had a bad year until the tax credits could be renewed for another two years in September 2004. This continuing market volatility heightens the risks to wind-turbine manufacturers and their customers, owners, and lenders, slowing industry development.

Other New Renewables

Other technologies in the new-renewable category are also being developed, including biomass, geothermal, and ocean-based energy. These technologies are deployed based on local resource availability and relative economics. In addition, numerous research initiatives attempting to develop exotic sources of power such as nuclear fusion are also being pursued.

Biomass As discussed in chapter 2, biomass in the form of wood for cooking and heating was the original fuel used by human beings and remained dominant until coal's ascension in the late nineteenth century. In many developing nations, most energy still comes from wood and other biomass sources, such as crop waste and animal dung. Worldwide, biomass supplies 9 to 13 percent of all energy, depending on the method of estimation.[32] This makes biomass a larger current contributor of total energy supply than either hydropower or nuclear energy.

Biomass energy is obtained from various forms of plant and animal-waste matter. Wood and dung are the established biomass leaders, but some industrial-scale projects seek to cultivate novel biomass crops such as hybrid poplars, hybrid willows, and switchgrass (all under development by the U.S. Department of Energy's Bioenergy Feedstock Development Program) for their energy content.[33] Most biomass energy is released by burning, but more efficient methods of capturing the stored photosynthetic energy in biomass are also being explored. These include biogasification—the process of using heat and catalytic chemistry to produce synthetic gas (*syngas*) from biomass, which often is used in efficient cogeneration plants that produce both heat and electricity. Bioenergy can also be harnessed from accelerated digestion of biomass by certain bacteria, which releases energy in the form of methane gas.

The real hope for biomass-based energy is that it can be harnessed to fill the future gap between the supply of and demand for transportation fuels. In generating transportation fuel from biomass, two approaches are generally used—creating liquid ethanol from the fermentation of plant matter and burning modified plant oil as a direct fuel (biodiesel). Brazil, for example, has focused on ethanol derived from sugar cane, developing and employing one of the world's most successful alternate-fuel programs over the last decade that now requires that all cars be able to burn both ethanol and gasoline. As a result, biomass currently represents 27 percent of Brazil's gross domestic primary energy production, with 40 percent of the cars in Brazil running on pure ethanol and the remainder using a blend that is 22 percent ethanol and 78 percent gasoline.[34] Europe, on the other hand, has led the way in biodiesel, cultivating large fields of rapeseed (canola) for this purpose, coloring areas of the European countryside a striking yellow every spring while at the same time marginally reducing Europe's reliance on imported petroleum.

Environmentally speaking, whenever biomass fuel can be derived from existing animal and plant wastes, it provides energy with little environmental impact. However, biomass in the form of wood burned for fuel is also generated from existing stocks of organic material embedded in the world's forests, resulting in deforestation and soil degradation in many tropical nations in Asia, Africa, and South America. It is also possible to cultivate dedicated biomass, such as being done in Brazil and Europe, on an industrial scale, but this requires a significant amount of land, water, and in some cases fertilizer and pesticides. Lack of necessary water and arable land will prohibit many countries that could benefit most from these technologies, such as China and India, from relying on biomass as a major contributor of future energy supply. In fact, outside of the United States, Canada, Brazil, and parts of Europe, few areas have sufficient agricultural capability to produce excess biomass for fuel, and none really has any capacity to produce quantities sufficient to export to less agrarian nations. As populations grow and natural resources are strained in the coming decades, the required inputs of soil and water required for this form of energy will be available in fewer and fewer countries, limiting biofuel's potential to replace dwindling stocks of oil in the long-term.

Geothermal Geothermal energy is, like biomass, an odd candidate for a "new" renewable energy source, as it has been used for heating living spaces and water for thousands of years. However, modern technology is enabling power generators to tap directly into the temperature differentials of water in geothermal reservoirs to create electricity at competitive rates, estimated by the World Bank to be between 2.5 and ten cents per kWh.[35] Geothermal plants run over 80 percent of the hours in a year, making them efficient users of their capital equipment and a good provider of base-load power, similar to nuclear power and likely cheaper where the geothermal resources are available. Many countries with natural sources of geothermal energy—such as the United States, Iceland, Mexico, Italy, Indonesia, and the Philippines—are beginning to harness this power, and the industry is estimated to have grown at about 7.5 percent per year between 2000 and 2005.[36]

The downside of geothermal electricity is that it still creates substantial risk for investors because the technology is relatively untested in the complete variety of locations and conditions in which it could be employed. Issues such as reservoir adequacy and depletion are not commonly or completely understood and can vary wildly from site to site. Also, industrial geothermal energy is not as environmentally friendly as it might first appear because it emits considerable amounts of gas, primarily in the form of sour and poisonous hydrogen sulfide gas and carbon dioxide, though considerably less than an equivalent amount of fossil-fuel use it may replace.[37]

Ultimately, deploying industrial-scale geothermal electricity as a primary source of base-load electricity is restricted by the location of viable sites. Geothermal power is economically viable only in countries or regions where geology happens to allow capture of the underground energy sources. The Geothermal Industry Association predicts that worldwide geothermal market potential is no more than 6.5 percent of all electricity generation.[38] The technology has good potential in certain locations such as Iceland, the Philippines, and possibly Bolivia and therefore is predicted to play a part in the energy mix for those countries. In most large industrialized nations, however, the future is limited.

Ocean Power Wave power, tidal power, and ocean thermal power are examples of marine energy technologies that are being developed to exploit the vast energy in oceans. Scientists are pursuing many creative approaches, such as using thermal energy differentials or the movement of water as tides and waves to create electricity. While some of these approaches show promise and research continues, thus far only France and Canada have any significant amount of ocean-based electricity generation in production, all in tidal power. Globally, all types of ocean power combined, including those in testing and pilot deployment, amount to less than 0.03 percent of total electricity-generation capacity.[39]

Owing to their reliability, ocean-based technologies may eventually become cost-effective for providing base-load electricity to populations near the sea, and certain types of ocean thermal electricity also have a by-product of desalinized water that could help to provide combined water and power to coastal cities. However, ocean-power technologies will always be affected by two problems. First, the issue of finding viable locations limits the size and scale of the potential generation market. Second, high maintenance costs result from placing sophisticated equipment in inaccessible locations such as the bottom of the ocean. While further research is merited, ocean power is not likely to become a widely used source of electricity without substantial improvements in reducing the cost and improving the science.

Fusion The technology for producing fusion energy is touted by scientists and industry supporters as a clean alternative to fission nuclear energy that results in dramatically reduced radioactive by-products and a virtually limitless supply of energy. However, despite supporters' claims in the 1950s that this type of power would be ubiquitous in fifty years, the International Thermonuclear Experimental Reactor (ITER) the leading international research consortium on fusion energy, claims that the realization of this promise is still fifty years away.[40] Fusion reactors have never provided the smallest level of steady power generation even under the most ideal conditions, despite U.S. government expenditures of over $10 billion on fusion research since the 1950s.[41]

Even so, ITER represents a multinational effort by the governments of the United States, Europe, Russia, China, Japan, and Korea to build a reactor to explore the scientific and economic merits of fusion at an expected cost of at least $10 billion.[42] The construction time of the test reactor (which will not actually generate electricity) is estimated at ten years, and the results from testing this device will theoretically allow for the design and construction of commercial-grade demonstration generators, with production-scale power plants available no sooner than 2050 under the current schedule.

The merit of using scarce resources for such a project is highly questionable economically. More effective ways of spending this research and development money include energy efficiency, emissions reduction, and other economically viable renewable-energy technologies that have more certain and immediate returns to their development. The reality is that there is very little chance that fusion technology can be developed, tested, made cost-effective, and deployed quickly enough to affect climate change or the world's current critical energy situation.

Diseconomies of Scale

One common hidden characteristic of nearly all of the technologies discussed in this chapter will limit their long-term economic viability—that is, their need to be deployed on a large scale to be cost-effective. To have any chance of being cost competitive in global energy markets, all of these technologies—hydro, nuclear, wind, and the other new renewables—must be developed on a utility scale in central facilities similar to existing fossil-fuel and nuclear plants. Yet the power usage of most homes and businesses is measurable on a much smaller scale—that is, in watts or kilowatts rather than megawatts or gigawatts. Centralized energy sources must rely on the existence of (and integration into) a utility grid for transmission and distribution to all nonindustrial users. Such systems are costly in all countries and unavailable in some, particularly in those developing countries that will experience the largest growth in energy demand in the coming decades.

The second half of this book explores the economic implications of this fundamental characteristic and its effect on a changing global energy mix that includes cost-effective distributed sources of energy and electricity. In the meantime, it is necessary to consider the role of hydrogen fuel cells, an emerging energy technology that is helping to bridge the gap between these new sources of electricity and the need for energy in transportation and thermal heating applications.

Decarbonization and the Hydrogen Promise

Several of the most promising new renewables suffer from the fundamental problem of intermittency: they produce electricity only when the wind blows, the waves roll, or the sun shines. As they are increasingly deployed, the costs and environmental advantages of these intermittent sources of electricity are somewhat mitigated by the need to maintain backup generator capacity that can meet fluctuations in supply—that is, that are *dispatchable*. In designing an electricity system that ultimately incorporates a significant proportion of intermittent sources, two potential strategies can be employed.

First, intermittent sources can be used only as a supplemental energy technology to defray the use of more traditional fuels when the intermittent energy source is available. Despite their intermittent availability, some of these intermittent sources are more predictable than others: tidal power is quite reliable, wind is not always so, and wave and solar somewhere in between. Depending on the way in which the predictability of the intermittent energy source corresponds to when energy is demanded, a significant minority of energy demand can be met by these intermittent sources.

Second, intermittent energy sources can be made more valuable in many more applications if a method of energy capture and storage is developed for use during those times when the original source is unavailable, creating the equivalent of stored battery power. While most attempts at energy storage have traditionally employed actual batteries, a number of large, industrial-scale technologies are being proposed to correct for intermittent technology or smooth out generation costs, including flywheels as well as pumped air and water. Another emerging solution to the energy-storage problem is the use of hydrogen and fuel cells as a method to store excess energy until it is needed. Unfortunately, many myths pervade the discussion about hydrogen-based energy technologies, masking the paths by which this technology will likely be adopted.

The hydrogen-power paradigm is the end result of a long process of fuel *decarbonization* that has been going on in varying degrees since the beginning of the human relationship with energy.[43] *Decarbonization* refers to the changing relative amounts of carbon and hydrogen in the fuels burned to generate energy, including wood, coal, oil, and natural gas. When each of these fuels combusts, both the carbon and hydrogen within them combine with oxygen and release energy in the form of heat. Only the carbon, however, causes undesired emission problems (such as air pollution and global warming) in the process of burning—specifically, the derived carbon dioxide and carbon monoxide. When hydrogen burns, it does not have the same pollution effect as carbon. Therefore, the less carbon available in the fuel being consumed, the more it is considered clean.

Wood is carbon intensive, containing about ten atoms of carbon for every atom of hydrogen. Coal, which began to replace wood at the onset of the industrial revolution, has one to two atoms of carbon for each part hydrogen. Oil has only about one carbon atom for every two hydrogen atoms and is regarded as a cleaner fuel than coal. The reputation of natural gas as a clean fuel stems from the fact that it has only one atom of carbon for every four atoms of hydrogen. Less carbon is released per unit of energy produced from natural gas than from any other fossil fuel. Although the world's total energy use and total carbon emissions have increased as energy use has shifted in emphasis from wood to coal to oil to natural gas, the ratio of carbon to hydrogen has steadily dropped. Creating energy via pure hydrogen is the final step; eliminating carbon completes the transition to clean fuel.

But where does hydrogen fuel come from? Hydrogen is by far the most common element in the universe, but it is not often found in H_2, its free form. Free-form hydrogen is currently harnessed by liberating it from larger hydrogen-containing molecules by one of several high-temperature chemical processes. About half (48 percent) of the free-form hydrogen produced today comes from *reforming* natural gas (reacting it with steam to produce carbon dioxide and hydrogen), while nearly all of the remainder comes from reformation of oil (30 percent) and coal (18 percent).[44] Currently, hydrogen is produced for two main industrial uses—to create ammonia that is primarily used to produce nitrogen-based fertilizers and to refine petroleum by hydrotreating it into its many useful forms, a process known as *cracking* the petroleum. The amount of hydrogen manufactured for these purposes has been growing annually at double-digit percentages for the last fifteen years or so, mainly from an increased demand for refining applications.[45]

The real promise of hydrogen, however, comes from its potential use in fuel cells. Fuel cells are devices that combine hydrogen with oxygen to generate electricity, producing heat and water as by-products (chemically expressed as $2H_2 + O_2 \rightarrow 2H_2O$ + energy). Fuel cells generate energy from hydrogen far more efficiently than simple burning does, and both of its by-products may be used in secondary applications, depending on how and where the fuel cell is otherwise configured. For example, excess heat from the reactive process can be used to heat buildings, and pure water has many practical cleansing, drinking, and irrigation applications. Because fuel cells lack moving parts, they are remarkably easy to maintain, which has the added benefit of increasing their lifespan and reliability while reducing operating costs. Finally, they are quiet, which makes siting near the locations they serve a viable option.

Both stationary applications for fuel cells in power generation and potential transportation applications are possible, with different fuel-cell configurations appropriate for different uses. For stationary fuel-cell generators, there are three potential markets. The first market for fuel cells is as backup generators. While this solution has proven cost-effective, growth in this market has been slowed by customers' lack of experience with the technology and their need for perceived reliability in stand-by power applications. The second market for stationary fuel cells is large-scale building power generation up to one megawatt. In this application, electricity and heat production from the fuel cell can be used, improving the overall fuel-to-useful-energy conversion efficiency to around 70 percent, about the same as the most efficient natural-gas generators. Historically, growth limits in this market have been a result of the large number of distributed generation alternatives for inexpensive power at this scale, including a variety of fossil-fuel generators such as natural-gas cogenerators. Recent increases in the cost of natural gas are meriting a reexamination of the comparative economics for fuel cells in these applications. The third potential market for stationary fuel-cell generators is residential applications, where they would need to provide approximately one kilowatt of electricity plus heat and hot water to power an average home in the industrial world. The market limitations of this application include its cost and the inefficiency of the low-temperature fuel cells that would be appropriate for in-home applications compared to larger, commercial-scale fuel cells. Potential markets for all of these applications exist and will continue to develop over the coming years, but the key applications do not yet have the critical mass to decrease costs and to drive continued market growth.

Fuel cells for vehicles are particularly exciting because of their perceived potential to end the dominance of the internal combustion engine in transportation but remain a long way from widespread use. Three basic problems continue to limit deployment of fuel cells for motive power—efficient hydrogen storage, fueling-station infrastructure, and cost. Hydrogen is the lightest element but takes up a lot of space in its gaseous form. Filled with hydrogen compressed to five-thousand pounds per square inch, a standard-size gas tank container could fuel a car for only two-hundred miles, a distance insufficient for modern driving patterns and approaching pressures deemed dangerous in a potential traffic accident. A number of solutions to this dilemma are currently under development, including the use of solid metal hydrides to hold the hydrogen as a sponge holds water, creating a high-energy density while reducing the pressure needed to store a given amount of hydrogen. Solid metal hydride hydrogen-storage systems are already on the market for stationary use in the United States and are in the road-test stage for vehicular applications.[46]

In addition, the vexing issue of where to supply vehicles with hydrogen remains a problem. A study presented in 2002 by the Transportation Technology Research and Development Center at Argonne National Laboratory estimated that building the hydrogen fueling-station infrastructure needed to service 40 percent of the U.S. light-duty fleet would likely cost over $500 billion.[47] This equation creates a quandary: who should invest in a hydrogen supply infrastructure before a sufficient number of hydrogen vehicles are on the road, and who will buy the hydrogen cars before the fueling infrastructure is in place? While this dilemma will begin to be mitigated by early corporate fleet applications and city transit services, this economic impediment remains a significant obstacle to the wider use of fuel cells in passenger automobiles.

The largest obstacle to using hydrogen fuel cells for broad-based transportation applications remains their prohibitive cost. A relatively large fuel cell is required to deliver the power needed to accelerate a vehicle, such as the seventy-five kW unit being tested for DaimlerChrysler's Mercedes A-class sedans. Such devices must become more cost-effective, probably on the order of a hundred times, before they can begin to compete with conventional engines. At the same time, the conventional engine market is evolving as well. Automakers today are rapidly deploying a new class of hybrid vehicles, including the Toyota Prius and Ford's Explorer hybrid, which won the award for truck of the year at the 2005 Detroit International Auto Show. These hybrids are dual gas-electric (or diesel-electric) to maximize fuel efficiency during operation. By using fuel on acceleration and at high speeds and electricity for cruising and low speeds, these cars cut fuel use to half or less that of comparable gasoline-

or diesel-only models.[48] Future fuel-cell cars are going to have to compete with these dramatically more fuel-efficient automobiles to break into the mass market. Automakers and governments believe that the technology will ultimately cross these hurdles and therefore are investing billions of dollars annually in development. Eventually, fuel-cell vehicles will likely find market acceptance, but the economic and capital investment challenges are daunting and may delay wide deployment for two to three decades.

Even as people are touting the clean hydrogen revolution, the promised environmental benefits of hydrogen fuel cells largely evaporate when the source of the hydrogen is considered. With 96 percent of the world's hydrogen coming from reforming fossil fuels, there is no clear-cut environmental benefit to making this transition to fuel-cell technology. Though the hydrogen itself is clean burning, its primary sources emit as much carbon as the direct use of the fuels that the hydrogen is supposed to replace. For example, in the manufacture of hydrogen by steam reforming of natural gas, presently the cheapest method, seven kilograms (kg) of carbon dioxide are produced for each kilogram of hydrogen.[49] Critics charge that the Bush administration's 2002 National Hydrogen Roadmap ignores this defect and touts hydrogen cars as pollution-free while planning to produce up to 90 percent of hydrogen from fossil fuels.[50] Ultimately, the long-term benefits of "clean" hydrogen can be realized only if it is originally obtained at the outset from nonpolluting sources.

A simple nonpolluting source of hydrogen does exist—*electrolysis,* the use of electricity to break water into hydrogen and oxygen. Electric current can be applied to water to separate elemental hydrogen and oxygen in a process that mirrors the one that fuel cells use to generate electricity. Instead of $2H_2 + O_2 \rightarrow 2H_2O$ + energy, the reaction becomes $2H_2O$ + energy $\rightarrow 2H_2 + O_2$. It is this simplicity and complementary relationship that will ultimately drive the hydrogen energy market. Electricity will be used, in effect, to charge the hydrogen with latent energy potential that can be run through a fuel cell as needed to recapture the electricity. Both processes involve energy losses, but they are at least as efficient as existing energy technologies and often are more so. Despite the fact that today the bulk of electricity comes from fossil-fuel sources, the eventual design of capturing electricity from renewable sources and storing it as hydrogen will complete the transition to a truly clean energy economy.

As the cost of natural gas inevitably rises and cheaper electricity from renewables becomes available, the relative cost of electrolytic hydrogen will drop, and the cost of electricity derived from reforming fossil fuels will rise. Once electrolytic hydrogen becomes the cheapest source of hydrogen, the central question will focus on the best location configuration for hydrogen creation. Specifically, two alternatives will arise. The hydrogen could be produced in bulk (using lower-cost industrial-scale electricity generation) and then transported to the locations where it is needed, such as homes, offices, and fueling stations. Alternatively, the hydrogen could be electrolysized on site where it is needed using either grid-based or locally generated electricity. The optimal choice will depend on the relative costs of electricity transmission and bulk hydrogen delivery and may differ by location and application. However, given the bulkiness of elemental hydrogen and the huge amount of specialized capital required to develop a delivery system for it, it is likely that local onsite generation will have an economic advantage, at least in retail-scale applications. With small buffer stocks of hydrogen supply, relative generation cost of hydrogen becomes the only relevant factor in deciding how to power fuel cells cost effectively.

The final stage of decarbonization is now within sight as the transition to new zero-carbon technologies is driven by the rising cost and decreasing availability of fossil fuels. Energy users are being forced to use existing supplies of energy more efficiently and change their methods of electricity generation and energy storage. Eventually, both stationary power and transportation power will be powered through a combination of renewable electricity and the hydrogen it creates, and using the same sources of energy and the same generation and storage methods will create substantial efficiencies, dramatically reducing the cost of the separate infrastructures for each type of application. However, a huge amount of renewable energy will need to be generated all over the world to meet increasing energy demand while simultaneously replacing dwindling use of traditional fossil fuels. The next chapter discusses the set of energy technologies that can be made ubiquitous and cheap enough to provide the needed amount of such energy.

5

Solar Energy

Many energy-industry observers consider solar energy a theoretically elegant but unrealistic solution to the imminent gap between global energy supply and demand. Everyone agrees that clean, limitless, free energy from the sky sounds ideal, but more practical considerations such as relative cost and the sheer scale of the current energy infrastructure seem to doom solar energy to follower status for years to come. Other sources of energy, both conventional and renewable (including wind, geothermal, and biomass), appear to be cheaper, easier to deploy, and better funded and currently enjoy popular support in the media and renewable-energy advocacy circles. In addition, memories of false starts and unfulfilled promises during the twentieth century have tempered general optimism about solar energy's potential. This credibility gap exists not only among members of the conventional energy industry—fuel providers, electric utilities, and all other interested parties—but also among a larger group of environmentalists and solar-energy system installers. Many of these people invested time and money to promote solar energy in response to the first OPEC oil shocks of the 1970s, only to be abandoned after 1982 by the national governments that had supported them. The memory of this disappointment lingers, promoting skepticism that solar could be a viable economic energy solution without substantial government subsidies.

Rapid changes in the photovoltaic industry, technology, and institutional players over the last decade have dramatically altered PV's economic viability, and fundamentally transformed the competitive landscape of the energy industry. Today, solar energy and photovoltaics comprise a global, multibillion-dollar industry providing cost-effective energy to millions of people worldwide in many large and growing markets. As with most technologies, the cost-benefit calculation varies by each potential user and application, making simple generalizations difficult. As a result, the largest remaining obstacle to continued adoption of solar energy is the lack of reliable and current information about its true economic characteristics. This chapter puts this growing global industry in perspective by highlighting its history—its roots, its driving forces and characteristics, the current state of its development, and methodologies for estimating how the cost of producing PV will change as the industry matures and grows.

Types of Solar Energy

Typically, an informed discussion about solar energy is limited by various and confusing notions of what the term *solar energy* actually describes. Broadly speaking, *solar energy* could be used to describe any phenomenon that is created by solar sources and harnessed in the form of energy, directly or indirectly—from photosynthesis to photovoltaics. Many of today's environmentalists use the term *solar energy* in its most comprehensive sense to include certain new renewable-energy technologies such as wind power and biomass, arguing that these sources derive energy from the sun, however indirectly. More conservative uses of the term, such as the one that this book employs, discuss direct-only solar sources, whether active, passive, thermal, or electric—that is, sources of energy that can be directly attributed to the light of the sun or the heat that sunlight generates.

This more restrictive classification is useful because a more general characterization of solar energy that includes wind and other technologies tends to obscure various isolated trends within the broader renewable-energy industry. Many renewable-energy technologies sometimes lumped under *solar energy* have very different economic characteristics, making it difficult to draw meaningful conclusions about them. Since the economic drivers discussed in the second half of this book do not apply to all technologies equally, it is helpful to be precise when analyzing specific industrial transformations and the markets in which they will occur.

Understanding direct solar energy requires examining three key continuums in the methods of harnessing it: (1) passive and active, (2) thermal and photovoltaic and (3) concentrating and nonconcentrating. Every solar-energy technology features some combination of these characteristics to harness sunlight. *Passive solar energy* requires a building design that is intended to capture the sun's heat and light. In passive solar design, heat and light are not converted to other forms of energy; they are simply collected. This is done through various design and building methods such as orienting a building toward the sun or including architectural features that absorb solar energy where it is needed and exclude it where it is not useful. The simplest conceptualization of passive solar-energy design for building is in a greenhouse, a design that allows solar light to pass into the interior and then captures the heat it generates inside to maintain year-round growing conditions. Passive solar features—some of which have been used in building design for thousands of years—include site selection and building placement that maximizes synchronized heating and lighting, windows placed in south-facing walls, vents and ducts moved to capture heat through the building, and *trombe walls* (dark, south-facing walls that absorb light and heat), wide eaves, heat-storing slabs, and superinsulation. Passive solar is an elegant way to harness the sun's energy, but it usually has to be designed into the original building plans to be made cost-effective. Once a building design has been finalized with siting, orientation, and structural elements, it is often prohibitively expensive to change or retrofit the facility to capture additional passive solar-energy benefits.

Active solar energy refers to the harnessing the sun's energy to store it or convert it for other applications. These applications include capturing heat for hot water that can be used for cooking, cleaning, heating, or purification; producing industrial heat for melting; or generating electricity directly or through steam turbines. The common characteristic is the active and intentional collection and redirection of the solar energy. These active solar solutions can be broadly grouped as either *thermal* or *photovoltaic* according to the method by which they generate energy for transfer or conversion into other useful forms. Thermal applications include all uses of the sun's energy in heat-driven mechanisms, such as heating water or some other conductive fluid, solar cooking and agricultural drying, or other industrial heat-collection applications—for processes as varied as water treatment or hydrogen generation through water decomposition. The most powerful solar thermal applications are used to superheat water and convert it to steam, which is then used to power a conventional steam engine for thermal electricity generation. Prior to the middle of the twentieth century, all industrial applications of solar energy were thermal in nature, and many of the simplest and most widely used remain so today, including the millions of rooftop solar water-heating systems installed around the world.

Solar photovoltaic is the state of the art in active solar electricity generation. By capturing the photonic energy of light on materials of a specific molecular structure, direct electric current is produced. The photoelectric effect (the description of which won Albert Einstein his Nobel Prize in physics in 1921 and which he believed to be more valuable than his work on the theory of relativity) allows an electric charge to be created on a semiconductive substrate that has been doped with chemical additives to create opposing positive and negative layers.[1] Photons of sunlight striking this surface facilitate an electron moving from the positively charged layer to the negative, creating an electrical current. This shifting of electrons in photovoltaic energy generation occurs without the need for moving parts and in proportion to the amount of light striking the surface. The useful lifetime of a photovoltaic cell is a function of manufacturing methods and the atomic stability of the substrate material, but some PV cells have been in operation for decades in space-based satellite applications, one of harshest possible environments. Land-based applications based on silicon material for PV cells are often warrantied by manufacturers for twenty-five years or more, although the expected useful life is much longer.

The final distinction in solar applications is *concentrating* and *nonconcentrating*. Concentrating solar applications use mirrors or lenses to focus sunlight. Concentration can significantly increase light intensity in the focus area, similar to the way in which a magnifying glass burns a hole in a leaf. Industrial-scale concentration can be achieved by the *trough method,* in which a long, troughlike parabolic mirror focuses sunlight along the length of a fluid-filled pipe suspended above the mirror. Large-scale concentration can also be created via an array of sun-tracking mirrors arranged to focus sunlight on a central point for thermal or photovoltaic use. Arrays of lenses can concentrate energy on photovoltaic cells, which tend to operate more efficiently (that is, convert more of the sunlight that strikes them into electricity) when the light is brighter. Concentrating systems are, by their nature, more complicated to build and manage than nonconcentrating systems and contain equipment with moving parts that suffer wear and tear as well as problems relating to the significant heat generated by them. Nonconcentrating systems, which allow the sunlight to fall on their energy-gathering parts without concentration by lenses or mirrors, are usually simpler and therefore less expensive maintain; however, they achieve correspondingly lower intensities and temperatures. Nonconcentrating systems include those that use direct sunlight to heat close-set pipes (as in a domestic hot-water system) or open water (such as a swimming pool), as well as most PV panels commonly seen on the roofs of houses and in stand-alone signs and lighting.

Figure 5.1 shows the breakdown of modern active forms of harnessing solar energy, both thermal and photovoltaic, among the various sizes of the generators used. The size classifications (centralized, large distributed, and small distributed) correspond to the different types of users that can use the amounts of power generated (utilities, commercial users, and residential users respectively). These distinctions will be examined in

Type of Solar Energy

System Size	PV	Thermal
Centralized (2 MW–GW)	Concentrating photovoltaic arrays (CPV) Utility-scale PV	Concentrating solar thermal power (CSP)
Large distributed (20 KW–2 MW)	Commercial-building PV	
On-site distributed (<20 KW)	Residential PV	Home solar hot-water systems

Figure 5.1
Today's mix of active solar-energy technologies by size and type.

later chapters to discuss the evolving economic decisions that each of these potential users of solar energy face.

A Short History of Solar Energy

Various forms of solar energy have been used since prehistoric times.[2] In fact, passive solar applications have been used in building design and construction for thousands of years.[3] Other early efforts included attempts to harness the power of the sun to wage war. Mythic stories abound regarding the third century BCE scientist and mathematician Archimedes, who, from the safety of the ramparts, defended ancient Syracuse from a Roman military invasion by using an array of solar mirrors to set fire to enemy ships in the harbor.[4] Though the accuracy of these stories is disputed by historians, there have been many recorded attempts to use lenses or mirrors to harness the power of the sun as part of strategic defense and warfare over the last fifteen hundred years. These experiments, none of which seem to have found much practical or long-lived use, included concentrating solar energy to burn, blind, or intimidate the enemy. Other proposed solar-energy applications were more commercial and industrial in nature. Leonardo Da Vinci, inspired by accounts of Archimedes' use of mirrors at Syracuse, designed a gigantic bowl mirror four miles across, to be built and used for large-scale industrial applications, including melting metals. Though the massive mirror was never built, Da Vinci foresaw, by many hundreds of years, the methods by which the sun's power would finally be harnessed for peaceful, commercial applications.[5]

Solar-energy technology saw a burst of new practical applications during the late-nineteenth-century industrial revolution, driven by three solar-energy inventors on different continents. This period in industrial history ushered in a grand expansion in knowledge and invention as entrepreneurs pursued many new energy technologies alternatives to wood and coal. As in all periods of rapid technological growth, an efficient form of commercial Darwinism determined winners and losers. Inventors and entrepreneurial businessmen developed new technologies. Society adopted those that operated and performed faster, better, and cheaper, while all others were put on the shelf until changing relative costs or technological breakthroughs created economic justifications to revisit them.

The first of the solar inventors was William Adams. In Bombay, India, in the 1860s and 1870s, Adams, a former British patent officer and engineer, conducted various solar-energy experiments and created practical devices such as a solar cooker to help ease energy shortfalls and depletion of local wood fuel in colonial India.[6] A solar cooker is effectively a large, bowl-shaped, concave mirror that, when pointed at the sun, creates a cooking hot spot by concentrating the sun's rays at a central focal point. Through trial and error, Adams determined that this technology was an effective means of heating and boiling water, though meat cooked in this way had a disagreeable flavor and smell, a problem Adams solved by using ultraviolet filters to block out the offending part of the solar spectrum. In addition, Adams created a solar-powered boiler for running a steam engine using a bank of flat mirrors to concentrate solar energy on a central vessel. Another of Adams's inventions concentrated solar energy to distill sea water into freshwater for the British enclave at Aden on the Red Sea. His groundbreaking efforts provided tangible evidence of potential solar-energy applications to many across the British empire.

At around the same time period, Augustin Mouchot, a French schoolteacher and inventor, attempted to develop solar-energy generators. He worked on the development of a solar cooker similar to that of Adams but also made and demonstrated a host of practical solar-energy devices, culminating in a large exhibition in 1880 in Paris, where he presented a variety of devices, including his large "Sun Engine," which operated a printing press on which he printed an edition of his newsletter *Journal Soleil*.[7] He also displayed a variety of solar cookers and, much to the amazement and delight of the crowd, a solar machine used to create ice.[8] Mouchot's state-of-the-art devices captured popular imagination, but it was his efforts to develop energy storage that defined his place in history. Mouchot was the first inventor who attempted to use the power of solar radiation to decompose water into its base elements of hydrogen and oxygen and then recombine them to generate electricity, much like the fuel-cell technology of today. While he felt that this was a potentially revolutionary solution to the problem of energy storage and the intermittency of the sun's availability, the increasing scale and cost advantage of coal-based energy prevented Mouchot from developing an economically viable solar-hydrogen energy solution.[9]

John Ericsson, the third solar inventor, was a Swede who moved to the United States in 1839, earning fame and fortune as the designer of the iron-clad Union ship the *Monitor,* which is credited with altering the course of the U.S. Civil War.[10] After the war, Ericsson turned his attention to solar energy and began extensive experiments in the 1870s that continued until his death in 1889. He developed a solar-power engine using hot air to run pistons, an efficient design that limited energy waste. Unfortunately, since the solar-powered machine could perform adequately only in direct summer sun, it remained only marginally valuable unless the problem of energy storage could be overcome.[11] Ericsson experimented with both compressed air and electric batteries in his pursuit of an effective and efficient energy-storage technology, but he was never able to find an economical solution to the problem of storing energy to work these engines in times of low or no sunlight. He did, however, make many significant contributions to the design of solar-energy collectors, including parabolic troughs, many of which are the basis of modern solar thermal designs.

Solar Goes Commercial

Solar energy's first chance for wider commercialization occurred between 1900 and 1915. Using the accumulated knowledge of earlier solar inventors, Aubrey Eneas, a solar entrepreneur in the American Southwest, developed larger solar collectors to power steam engines and pumps for agricultural irrigation water. The dry deserts of this region made an ideal test market with plenty of sun and few alternative fuel sources. Residents of the region were also already comfortable with the practical use of solar energy, with some 30 percent of the homes in Pasadena, California, using solar water heaters to generate hot water by around 1900.[12] Eneas, in an attempt to create commercial-scale solar-energy applications, designed and built a large truncated cone collector to superheat water that powered a steam engine for running irrigation pumps. He managed to convince several customers of the merits of his design and installed a handful of these systems across California and Arizona. Ultimately, however, the construction methods of these large cone collectors proved susceptible to strong and unpredictable weather of the desert region including dust devils, wind storms, and hail. Eneas eventually abandoned his attempts to commercialize concentrating solar energy, believing that they would never be economically viable.[13]

The last major pioneer to attempt to commercialize the power of the sun early in the twentieth century was another American named Frank Shuman. At the time that he began his foray into solar energy, Shuman was already famous for many useful inventions, including shatter-proof wire glass and the safety glass found in today's automobiles. His resulting financial success afforded him the time and money to pursue commercializing solar energy. Befitting someone who had made his fortune in the glass business, Shuman's first attempt to capture solar energy was through the use of a hotbox, a device similar to a greenhouse in which sunlight enters the box through a glass pane, trapping the heat and causing a dramatic temperature increase inside. By stringing a series of these hotboxes together, Shuman created enough useful heat to run a small boiler for a steam engine. While this design was simpler and less expensive than using mirrors or lenses to concentrate the sun's rays, Shuman ultimately decided to add a row of mirrors to each side to increase the heat generation. After perfecting his techniques in Philadelphia, he found investors and signed a customer in the British protectorate of Egypt for whom he delivered solar pumps for irrigation. Against his better judgment, Shuman was coerced by his partners to abandon the hotbox technology in favor of parabolic troughs similar to John Ericsson's original design. In the end, Shuman's solar-energy irrigation solution was substantially cheaper to operate than any alternative solution but was twice as expensive to set up as the next best solution, the coal-fired engine. Even though the up-front cost differential could be repaid from reduced operating costs in the first two years, Shuman's parabolic troughs remained a marginal choice for investors because there was little long-term financing available at the time.[14] The Great War of 1914 to 1918 forced Shuman to shut down his Egyptian venture, and he returned to America. For over half a century afterward, easy access to coal in the industrial economies of Europe and America along with the increasing availability of petroleum as a cheap power source for motive applications eclipsed nearly every attempt to commercialize the power of the sun.

Solar's Second Chance

Solar energy's next chance for widespread market acceptance came with the development of the first official photovoltaic cell in 1954 by three researchers at Bell Labs. Though the basic photovoltaic effect had been understood in the nineteenth century by Edmund Becquerel, the Bell Lab researchers' work in semiconductors and some fortunate laboratory accidents led to their development of the first working PV module. These modules quickly progressed by the end of the 1950s, stimulating a tremendous amount of excitement among research labs and the governments that funded them about the future of this technology.[15] Continued work and development in the 1960s, much of which was performed to support critical power systems for applications in satellites and space-based vehicles, led to increasing optimism by the international scientific community about the future of PV technologies.

The oil shocks of the 1970s provided a further boost for solar technologies, both PV and thermal. Supply fears and the rapid rise in fuel prices they caused led to strong government promotion of a variety of alternative-energy technologies, including solar energy. In the 1970s, the U.S. government established the Solar Energies Research Institute to help develop solar-energy technologies. President Carter's administration helped bolster the industry by approving a $3 billion program for the development of solar-energy technology and installing a showcase solar water heater at the White House. Excited about the potential for clean, independent energy sources, many American citizens and entrepreneurs began to invest substantial time and money to promote these changes by developing businesses and installing solar water heaters and PV panels. This momentum was cut short with the changeover to Ronald Reagan's administration beginning in 1981. By 1986, the Reagan government had dramatically reduced funding for solar-energy research programs, reduced the federal tax credits for solar water heating, and removed Carter's showcase system from the White House roof. The message to the renewable- and solar-energy communities was clear, and the industry came to a virtual halt not only in America but throughout the world. America represented almost 80 percent of the world market for solar energy at that time, and when research funds in America dried up, the remaining governments of the industrialized world followed in step. A period of decreasing oil prices in the 1980s and early 1990s further diminished the perceived urgency to cultivate renewable-fuel options, leading to another two decades of inertia for the global solar-energy industry. Disparate inventors, research universities, and state energy agencies continued funding research and development in photovoltaic technologies, even as giant energy companies such as Mobil, Shell, and BP purchased the assets and patents of many of the original solar-energy technology companies.

Solar Comes of Age

Over the last ten years and underneath the radar within the broader conventional and renewable-energy industry, solar energy has emerged to make a third attempt at mass commercialization. This time, the opportunity is due almost exclusively to the efforts and programs of the national governments of Japan and Germany, which have led the way in promoting these industries. Although these two countries do not have a naturally high amount of sunlight, their lack of alternative fuel sources has created a dependence on expensive external sources of energy and therefore motivated them to develop less expensive, local, and renewable-energy alternatives. Japan's sunshine program and Germany's 100,000 solar roofs program, which have used various types of subsidies to stimulate robust domestic solar-energy industries, now account for 69 percent of the world market for PV.[16] In these markets and geographics, the current renaissance of interest in solar energy is finding its opportunity, and the cost reductions these markets have experienced have stimulated surprisingly powerful momentum and growth in the last decade. With the global PV market growing from 85 MW in 1995 to over 1.1 GW in 2005 (a 29 percent annual growth rate), the cost of producing PV systems has dropped from $11 per watt to as low as $5 per watt over the same period and continues to fall by 5 to 6 percent per annum.[17]

The Dawn of a Solar Industry

Photovoltaics can be applied cost-effectively at any scale, from handheld gadgets to utility-scale generation.[18] Each application has a unique character, but there are natural groupings. The first and most familiar type is using solar cells without any kind of battery solution and usually at small scale. Applications of this type include solar calculators, irrigation pumps, and freshwater distillers that operate only while the sun is shining. More complex systems include those that store excess photovoltaic electricity in batteries for use at night. For example, homeowners can buy photovoltaic yard lights that charge a battery during the day and glow after sundown until the battery runs out. Area lighting, highway construction signage, and roadside emergency phones are other examples. Together, these two types of PV application—small-scale, single-function systems with or without battery storage—comprise about 15 percent of the PV capacity installed annually worldwide.[19]

Another primary market for solar cells is providing power to homes that do not have access to electricity in other forms. The economics of these off-grid applications are compelling and are discussed more fully in the following chapters but solar photovoltaics are often a cost-effective solution in the absence of alternative off-grid energy solutions. Unlike gasoline-powered generators, which are the traditional alternative for off-grid power, solar cells require no fuel deliveries, operate silently, and never refuse to start. The system equipment includes the solar modules themselves as well as some form of mounting that can be either stationary or can track the sun on one or two axes to maximize the sun collected. Solar cells also typically require a lead-acid battery storage solution to provide power in hours when the sun is unavailable. A specific form of lead-acid battery called a *deep-cycle battery* usually works best for these applications because they last longer than traditional twelve-volt car batteries. These off-grid solar applications can be found in industrialized countries where there homes and businesses are located outside the range of the existing grid but more commonly are found in the developing world where access to reliable and consistent grid power is not available. Combined, these off-grid applications represent some 18 percent of the total PV installed worldwide on an annual basis.[20]

The bulk of the remaining solar market represents the growth of *grid-tied systems* for residential and commercial customers.[21] These systems use sets of PV panels ranging from a few hundred watts to a few megawatts of peak capacity and are located on rooftops of homes and buildings. During the day, as the energy collected by these PV systems exceeds the energy needed for the local homes or business, the system feeds the excess back into the utility grid at effectively the same rate that the customer would pay for that electricity, a concept commonly referred to in the United States as *net metering*. This system obviates the need for a localized battery solution for these installations because the utility grid is used instead as the equivalent of a huge storage battery. Once a building owner meets the connection requirements of the local utility, whereby the utility confirms the safety of the equipment and the correct connection parameters, a customer's meter can flow in both directions—positive at night when the home or business is using more power than it is generating and negative in the sunny part of the day when the opposite is true. This system does not require a battery but needs a device called a *grid tied inverter*. These inverters collect the direct current (DC) of solar PV and convert it into the alternating-current (AC) power of grid electricity. The remaining solar modules and associated mounting and wiring are essentially the same as any other off-grid application.

This grid-connected market is the engine driving the solar industry's commercial-scale growth and transformation. While total solar applications have collectively grown 29 percent per annum over the decade through 2005, the grid-connected segment has experienced a growth rate of over 50 percent per annum over that same period.[22] Figure 5.2 shows how much the growth of this segment has contributed to the total growth of solar photovoltaics. This sustained growth in the use of

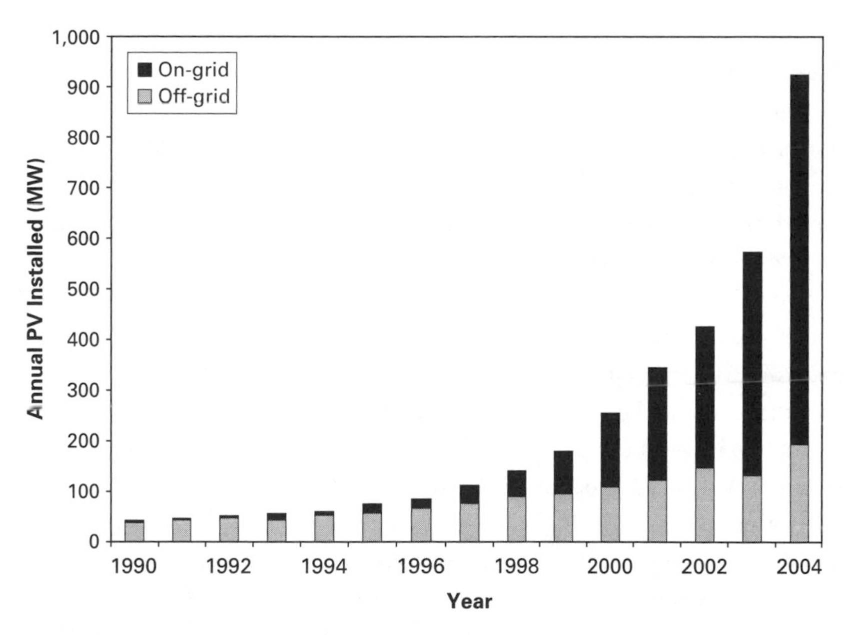

Figure 5.2
Growth in annual installations of grid-connected and off-grid photovoltaic cells, 1990 to 2004 (MW of peak capacity).
Source: Solarbuzz (2005).

grid-tied systems will continue to propel the PV industry in the coming decades as grid-tied PV economics and the technology's innate reliability increasingly provide incentives for customers to adopt such systems.

Finally, an application known as *centralized systems,* while small today at only 2 percent of the total PV market, should become increasingly relevant in the future as more large energy users and utilities adopt PV.[23] Using this application, industrial customers or utilities can take advantage of good solar characteristics in a given location to generate utility-scale power in large fields of ground-mounted solar arrays. To date, this application for PV is not widely cost-effective in the face of industrial-scale alternatives, though the next chapter shows where it is becoming so. Centralized systems have the potential to contribute significant amounts of electricity, as one-third of the earth's surface is covered by sun-rich deserts, creating a potentially vast amount of energy resource. Some 4 percent of which (just over 1 percent or the total land area of the world) would meet the entire world's energy needs from these sources even at today's efficiency levels.[24]

Geographic Markets

The driver of the solar electricity industry has been grid-tied applications but primarily from only a few geographic markets. As recently as 1998, the United States was the world leader in PV installations, but concerted government programs in both Germany and Japan have enabled these countries to surpass the United States in terms of PV capacity installed. Figure 5.3 shows just how quickly the markets have taken off since 2000 and projected growth in 2005 and 2006.[25] However, U.S. markets are beginning to develop further as many state-level governments institute programs to develop renewable-energy industries or markets.

The Japanese photovoltaic market is currently the largest globally, with over 1 GW of peak capacity installed, of which 277 MW were installed in 2004 alone, 36 percent higher than the prior year.[26] Over the last decade, this rapid growth resulted primarily from the Japanese government's residential PV-dissemination program, which created targeted incentives for installation of solar photovoltaic modules on residential rooftops. Chapter 9 details this successful government program to foster technology growth. The most important fact about this program is that it has helped lower the cost of installation of grid-tied residential PV in Japan by half since its

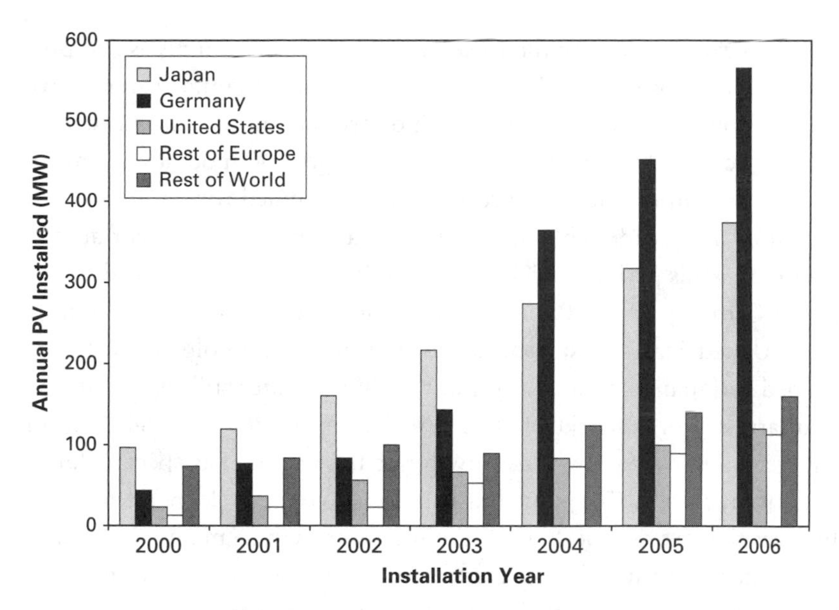

Figure 5.3
Growth in annual installations of photovoltaic cells in the major geographic markets, 2000 to 2004 with estimates for 2005 and 2006.
Source: Solarbuzz (2005).

implementation in 1996.[27] As of 2004, although the subsidy was reduced to around 7 percent of the system price from an initial level of 50 percent, demand continued to soar, signaling a belief by homeowners that the solution is now nearly at parity with retail electricity rates of twenty-one cents per kWh.[28] The government is forecasting continued robust growth even without the aid of ongoing subsidy programs and hopes to be adding well over a GW per year by the end of the decade.

The European market for photovoltaics, while significantly smaller than its wind-power market, has also seen dramatic growth over the last decade with 50 percent annual increases in domestic production in both 2003 and 2004.[29] Performance-based incentives to pay system owners a premium for generating clean PV electricity instituted in Germany in 1999 and renewed in 2004 have fostered PV growth to 366 MW in 2004, an amazing growth of 150 percent over the prior year, primarily in the grid-tied market.[30] Today, Germany represents 80 percent of the solar photovoltaics installed in Europe, though this percentage will decrease in the next few years as other countries such as Spain and

Portugal begin to increase their use of PV.[31] However, supply is not projected to keep up with demand because many manufacturers have already committed over a year's worth of production to customers, with Germany currently importing some half of its PV modules installed. Market expectations across Europe are for continued robust growth for the rest of the decade with some forecasts as high as 40 percent annual growth over this period, which would result in over 1.3 GW installed per year in Germany by 2010, out of a total of 1.7 GW for all of Europe.[32]

The United States' historical leadership in PV technologies has been usurped by Japan and Germany, but the United States still represents the third-largest world market.[33] In 2004, U.S. users installed 84 MW of peak generating capacity.[34] Its growth rate has not been as spectacular as those of Japan and Germany, but it has still averaged 25 percent annually over the last decade.[35] With the recent growth in many states' subsidy payments for installing PV systems, reimbursing up to half or more of the system cost, grid-tied residential and commercial systems have grown much faster. California has experienced the greatest growth in photovoltaic installations, today making up over 80 percent of the U.S. total.[36] In addition, the United States exports a net amount of 58 MW of solar cells to make up for supply shortfalls in Germany and the rest of the world, totaling 139 MW of PV cell production.[37]

Outside of these three main markets, which together comprise 89 percent of the world's total production of photovoltaic cells, the rest of the world produces 124 MW of product, 38 percent higher than in 2003.[38] These markets have experienced volume growth over the last decade similar to that in the Japanese and German grid-tied markets, averaging over 50 percent annual growth between 2000 and 2004, albeit from a low base.[39] Despite the relative benefits that this technology can bring to the developing world, there are limits to its adoption, such as lack of system financing as well as sales and service channels, which are explored in detail in the second half of this book.

Producers

In general, producers of these photovoltaic modules are located in the major markets in which they sell their products. Table 5.1 shows the top ten producers worldwide and their annual production for the last four

Table 5.1
The top ten global solar-cell producers in 2004 and their past production from 2001. (Production in peak MW.)

	2001	2002	2003	2004
Sharp (Japan)	75	123	198	324
Kyocera (Japan)	54	60	72	105
BP Solar (United States)	54.2	73.8	70.2	85
Mitsubishi (Japan)	14	24	40	75
Q-cells (Germany)	N.A.	N.A.	28	75
Shell Solar (Germany)	39	57.5	73	72
Sanyo (Japan)	19	35	35	65
Schott Solar (Germany)	23	29.5	42	63
Isofoton (Spain)	18	27.4	35.2	53.3
Motech (Taiwan)	N.A.	N.A.	N.A.	35
Total	**296.2**	**430.2**	**593.4**	**952.3**

Note: N.A. = Data are not available.
Source: Maycock (2005).

years. Given the relative size of the Japanese market, it is no surprise that four of the ten largest PV producers are Japanese firms and represent some of Japan's largest and most powerful industrial firms such as Sharp, Kyocera, Mitsubishi, and Sanyo. These companies are dominant electronics firms with market penetration and expertise in cost reduction via large-scale production development, and they have been primarily responsible for the reductions in system prices for these markets since 1994. In Europe, PV production is led by domestic producers such as Germany's Q-cells and Schott Solar and Isofoton of Spain and is also bolstered by BP Solar (British) and Shell Solar (Dutch). America's PV production is dominated by these two oil-company owned solar divisions, and General Electric entered the fray in 2004 by purchasing America's largest independent PV producer, Astropower, out of bankruptcy. Each of these major producers is looking at substantial additions to capacity in the next twelve months, with some planning to double their line capacity. Many PV producers outside of these major markets are also planning production increases, including Motech of Taiwan in the top ten global producers. Countries like China and India are leading the way in setting ambitious growth targets for solar photovoltaic installations.

Technologies

There is a constant and growing effort to find the most inexpensive and reliable technologies for producing these solar cells to continue to develop the photovoltaic market worldwide. While the dominant technology today is silicon based, many advanced silicon and nonsilicon variants are being explored to meet these needs.

Monocrystalline silicon cells are roughly similar to those PV cells originally created in 1954 at Bell Labs. Today, they are generally formed from ingots of pure silicon that are sliced into thin wafers and then chemically treated and etched to operate as solar cells in a process similar to that used to produce chips in the microprocessor industry. Their advantage is that they possess the highest levels of conversion efficiency for turning sunlight into energy. Their disadvantage is that they are the most costly to produce of all possible choices because they are fabricated using energy- and capital-intensive methods derived from similar processes in their microprocessor counterparts and often at a quality standard much higher than necessary for use in current photovoltaic applications. The alternative silicon-based cells are *polycrystalline* and are produced using slightly different manufacturing methods, creating a less efficient but also less costly end product. The major PV producers use variants of these technologies, with the American producers favoring the monocrystalline and the Japanese favoring polycrystalline. Combined, these basic silicon solutions make up 85 percent of the solar cells produced today.[40]

The remaining market for PV is comprised of a second-generation technology called *thin-film PV* that eliminates the need to have freely supported solar cells in favor of depositing the photovoltaic layers directly on a supporting substrate. This allows for a further reduction in cost and a more creative configuration of cells such as embedding the cells within building materials, rolled sheets, and roof tiles. The modules developed using this process are usually of lower efficiency and lower cost but have continued to improve in both respects as they are commercialized. With so many interesting applications being pursued, including those that eliminate the use of silicon entirely, most of the major industry research associations expect thin-film PV to be a major contributor to long-term growth of the industry as cost effectiveness improves. Ultimately, the progress of thin-film PV will be a function of how quickly technologies in silicon-based crystalline solutions reduce manufacturing costs and the third-generation technologies discussed below are deployed.

Third-generation (or 3G) technologies—so termed by Martin Green of University of New South Wales, who is one of the world's leading researchers on solar cells—are poised to make a much larger long-term impact in solar cells but probably not until the second decade of this century when these technologies will finally hit the mass market.[41] The technologies being explored include photosynthetic chemical processes similar to those that occur in plants and trees. There are also a number of exciting technologies for producing spherical solar cells—that is, small bubbles that reduce the amount of silicon needed, resulting in thin, flexible panels that can be used in, for example, microelectronics applications. In addition, about half a dozen companies are attempting to figure out ways to use printing technology to print solar cells directly onto a substrate, a development that could dramatically reduce the cost of manufacturing. None of these technologies is yet widely available, but they have the potential of bringing the cost of solar energy well below today's price once the technical hurdles of mass producing them are overcome.

PV Supply Chain

The modern PV industry is a $10 billion dollar industry worldwide and comprises many types of manufacturers, installers, and service providers. Figure 5.4 shows the PV production chain for the support services required to develop and deploy PV systems.

The solar-component manufacturers comprise the cell producers discussed above as well as the raw silicon and ingot producers, production-line equipment manufacturers, and providers of the other metals and raw materials needed to make PV cells. Product manufacturers (often but not always the same companies that make the PV cells) combine the cells with glass, frames, and electronic busing to create finished modules. These are then aggregated by distributors or installers with the balance of system components, including inverters, batteries when necessary, and wiring to be installed in end-user applications.

Supporting this process are the research and development by companies and research institutes; the banks and capital markets that provide

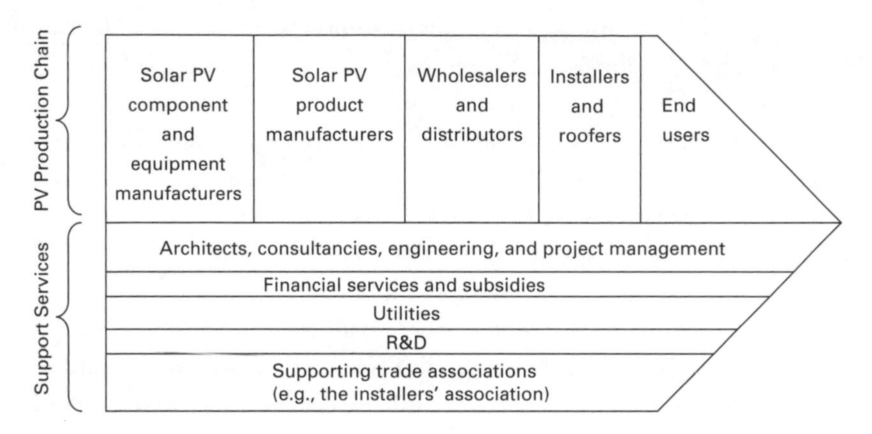

Figure 5.4
Photovoltaic industry production chain and various support services.
Source: Solarplaza.com.

the capital to create the manufacturing and finance the end-user installation of the systems; and the designers, architects, and engineers who enable integration of these systems in homes, buildings, and new centralized plants. The many phases of production require various technical skills and combine to make PV production a labor-intensive process among electricity technologies, even at an equivalent level of electricity production. As the industry grows, new jobs at all stages of the PV supply chain will need to be filled in proportion to the overall industry growth rate, with many of these at the local design and installation levels.

Trends and Projections in the Cost of PV

Typically, emerging technologies in any industry start off as expensive, complex, and inefficient to produce. The most promising ones—those with the potential to fill a previously unmet need or to provide an existing product or service more cheaply or effectively—begin to generate growing interest, and more versions and refinements are introduced. In this research and development phase, industry works out methods by which the technology can be produced and begins to envision its final form and application. As users are convinced of the technology's viability and long-term economic value, industry producers begin to produce limited quantities of standardized products to meet specific market applications. Mass production usually begins in limited lot sizes, but as market demand increases, producers usually find faster, better, and cheaper ways to produce their products. Sometimes just making more units can reduce the average cost of each unit because of bulk pricing on raw materials and a fuller utilization of people and manufacturing facilities, known as *economies of scale*. Institutional experience becomes embedded, too, as more efficient production methods are systematically discovered and reduce manufacturing costs.

The current cost of any technology, then, is partly a function of the history and scale of the production of that technology. A useful depiction of this process is the *experience curve*. This curve, often represented graphically, describes the relationship of the technology's current cost to the cumulative quantity produced over its history. In the case of an electricity-generating technology such as PV, cumulative quantity produced is measured in peak capacity, usually watts, produced. Therefore, experience shows up as decreasing cost as cumulative production increases.

In the case of photovoltaic modules, the cost to produce them in the late 1970s was around $25 per watt but has since dropped to less than $3.50 per kW, an 86 percent reduction, while the cumulated production has grown a thousandfold, or roughly ten doublings over that time. Calculating experience curve for grid-tied solar photovoltaic modules over the period from 1980 to 2003 implies a learning rate of around 18 percent, meaning that for every doubling in the installed volume of the technology over time the per-unit cost drops another 18 percent.

Forecasting future learning rates is as treacherous as forecasting anything else, but a few general parameters are helpful. Starting with a historical rate is fine, but that rate may change in the future, either increasing or decreasing. Typically, a technology experiences a slowing of its learning rate as it grows and becomes established. Because the most obvious and easy gains in cost efficiency are usually the first to be captured, driving costs out of the production process gets progressively harder as it matures. Scale of production helps further, but these gains also level off as an industry reaches critical mass. As a result, most technologies have a natural limit beyond which cost improvements slow to nearly nothing. Each industry is different, but at some point, usually well before market saturation, a technology stops gaining cost advantages from increases in scale. That being said, occasional breakthroughs in technology can occur that can greatly accelerate the learning rate for a period of time.

Once an experience curve is projected, the expected market growth of the technology in question can be layered in, and an expected annual cost reduction for the technology can be calculated. In the photovoltaic case, an 18 percent learning rate and a 30 percent annual market growth rate would translate into a cost decline of 5 percent to 6 percent a year—numbers that correspond with recent historical experience for both growth and cost reduction in the global PV module market.

Using assumptions for experience rates and total market growth, it is possible to build a forecast model of the expected future cost of PV electricity. Figure 5.5 shows a forecast model through 2040 for three locations with different levels of available sun and the expected worldwide growth of the underlying market.[42] As the next chapters explain, potential residential, commercial, and utility adopters of PV face different economic choices and different methods of financing the cost of their system. The underlying market growth rates span all of these sectors as it is cumulative global growth that determines expected system prices. The projected PV electricity costs in this forecast were developed assuming

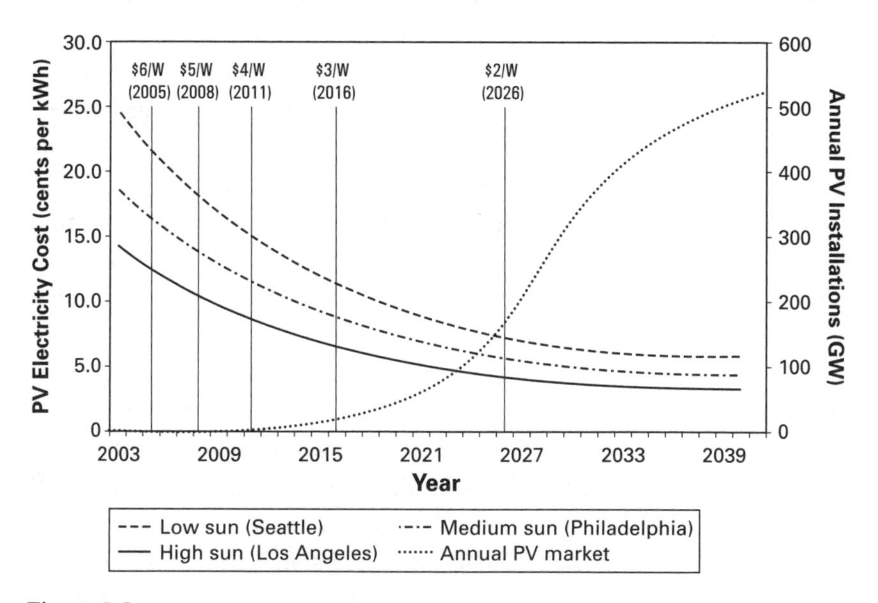

Figure 5.5
Forecast of the global annual market for PV, expected cost per watt of installed PV systems, and resulting cost per kilowatt hour of PV electricity (unsubsidized) for three sun scenarios through 2040.

the economics faced by residential customers in the United States and the type of supplemental mortgage financing that they would most likely use when installing PV on their home.

The projection model in figure 5.5 illustrates that, even under conservative assumptions, the evolving economics of PV will make these systems progressively cheaper over time, which will make PV electricity increasingly cost-effective in locations with a wide disparity of solar resources and allow PV to grow in dominance within the global energy and electricity mix. The cost levels of PV systems (in dollars per watt) are used for more specific forecasting later in the book and examined for reasonableness and limitations.

When asked about their views on various energy technologies, over 90 percent of people believe that solar energy is a desirable solution, making it the most popular of all energy technologies.[43] Although most people have never seen it deployed successfully, solar energy is popular because it is conceptually simple—almost deceptively so. Solar energy is safe and clean and has no moving parts, making it reliable and long-lived, and the sun as a source of energy cannot be bought, sold, or metered. As a result, solar energy offers nations and individuals unprecedented and unlimited control over their own vital source of energy. For these potential benefits alone, many people genuinely want to see solar energy become a widely deployed energy alternative.

Despite wide pessimism about the cost and capacity of PV technology, a dramatic growth in the rise of solar photovoltaic electricity over the last ten years has occurred primarily as a result of Japanese and German government policies that promote the development of grid-tied PV systems. As the next two chapters show, the increasing scale of global PV deployment has decreased the cost of grid-tied systems to the point where PV is cost effective in many large and growing industrial nations. As PV continues to be adopted in these markets and global production volume increases, predictable cost reductions will make photovoltaics more affordable for both industrial and developing-country consumers worldwide—leading to greater sales, larger volumes, and further cost reductions. The next section of this book explores specifically how and when these trends will shape the future of the energy industry.

III

Future Transformations

6

Modern Electric Utility Economics

As it has throughout most of our history, the demand for the services that energy and electricity provide is likely to continue to grow in a global economy that encompasses hundreds of millions of industrialized consumers who have expectations of future prosperity and billions who aspire to industrialized levels of prosperity in developing countries. People will continue to demand more lights, more cars and trucks, and more computers along with other modern amenities. To meet these needs, existing sources of energy will have to be spread further and used more efficiently, and additional sources of energy will have to be deployed. Effective and coordinated government policy might be able to meet this rising demand by aggressively pursuing efficiency improvements and disseminating "best practices" of energy generation and use to growing nations and economies. However, we do not live in a world of effective, coordinated government policy, particularly when resource hunger drives short-term decision making. Historically, improvements in the efficiency of energy use have not been able to stop the need to acquire new energy sources to satisfy increasing demand, and it will not likely do so in the future either.

Currently, the modern world almost totally depends on the stored solar energy embedded within fossil fuels for transportation, heat, and electricity. In the broadest sense, modern industrial capacity has been created specifically by and for the exploitation of this form of fuel-based energy, and all business-as-usual forecasts implicitly assume that the global economy will continue to be fueled in a reliable and cost-effective way. Nevertheless, the world's oil and natural-gas deposits cannot provide a constant energy output over the next few decades, much less an increasing one. Coal is the only remaining fossil fuel available to supplement the difference, but the environmental impacts of current coal energy-generation technologies are generally considered unacceptable. As the need for industrial energy continues to outpace supply, the inelastic nature of demand for these energy resources will drive increases in the prices of gas, oil, and coal. For both environmental and economic reasons, large quantities of the alternative sources of energy described in the last two chapters will need to be harnessed to make up the difference.

When selecting additional sources of energy to use in electricity generation, utilities have a portfolio of choices including nuclear energy, wind power, and solar energy. An examination of the costs of generating electricity by each of these methods as well as the type of power they provide shows that solar energy can already cost-effectively supply a portion of utilities' needs for daytime electricity, currently the most expensive form of electricity for utilities to produce. Using the tool of experience curves discussed in the last chapter shows that the relative competitiveness of solar electricity for utility-scale generation will continue to grow.

Forecasting Future Energy Prices

Many ways of forecasting energy supply and demand are used today by governments, research labs, and consultants. These range from highly technical economic models to scenario analysis of possible future worlds. Some of the most widely used figures come from the U.S. Department of Energy's Energy Information Agency (EIA), which publishes an annual forecast of world energy use—the International Energy Outlook (IEO)—and a similar projection that includes expected future world prices of fossil fuel and electricity—the Annual Energy Outlook (AEO).

The EIA projects a number of future scenarios in these reports, but the one cited most often is the Reference Case, the one asserted by the EIA to be the most likely to occur.[1] The 2005 IEO Reference Case's forecast to 2025 for global energy forecasts that world energy demand will grow at 2 percent per year from 411 quadrillion British thermal units (Btu) in 2002 to 644 quadrillion Btu in 2025.[2] This projected increase in energy supply is forecast to be met by all energy sources—coal, natural gas, oil, and nuclear energy. Under this forecast, natural gas will grow faster than all other fuels, supplying a higher share of energy demand in 2025 than today, and nuclear will grow the slowest, as many plants constructed in the 1970s and 1980s are decommissioned over this period when they reach the end of their forty- to fifty-year useful life.

Figure 6.1 shows the breakdown of projected energy demand by region, divided into industrialized nations, developing nations, and the former Soviet Union. As the graph shows, the bulk of the growth in energy demand will come from the developing world (particularly in China and India, which together contain more than one-third of the world's population) as rapid industrialization absorbs tremendous amounts of additional energy.[3] However, the industrialized world and countries of the former Soviet Union are also expected to experience increased energy demand.

The EIA updates its forecasts annually and compares current-year forecasts to those of prior years to show the trends in its own projections. However, with something that has the complexity and long time horizon of global energy supply and demand, it is necessary to look at the results of any forecast to see if the combined outputs of the projection models can be realistically achieved. Doing this for the IEO projections,

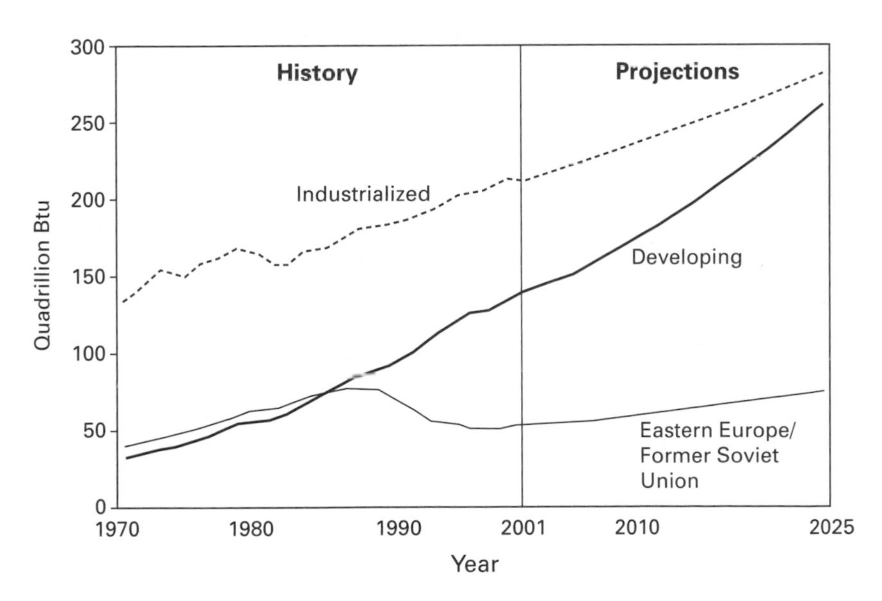

Figure 6.1
World energy consumption by region from 1970 to 2001 with forecasts through 2025 (quadrillion Btus).
Source: EIA (2004).

the outputs of the EIA's model appear to be overly optimistic and mutually inconsistent in terms of prices and volumes of energy used. EIA's IEO model shows a world that by 2025 has seen a 54 percent increase in total energy supply from 2002. In the AEO, the EIA projects that oil, natural gas, coal, and electricity will be roughly the same price in real terms in the year 2025 as they were in 2003.[4] The model thus assumes that through some combination of productivity improvements, additional production, new pipelines, new technology, and market forces, the world can increase energy output by over half through 2025 while keeping prices fairly stable—a scenario likely to prove drastically incorrect for several reasons.

For reasons outlined in chapter 3, by 2025 the world will almost certainly be well past peak production of oil in nearly every non-OPEC country and quite possibly in most OPEC countries as well. Even allowing for more optimistic views of the amount of oil available, a significant minority of the world's oil fields are at or beyond their peak production today and will experience decreasing oil output over the next twenty years. At a minimum, therefore, satisfying increased oil consumption will require larger output from the remaining producers. The EIA forecast assumes that output will increase from around 78.2 million barrels of oil per day in 2002 to nearly 119 million barrels per day in 2025 in the face of an undisputed decline in production by many producers, including the United States.[5]

Next, even if the necessary oil reserves are available, the production and delivery infrastructure may be inadequate. To bring these additional 40 million barrels a day to market in 2025, all links in the supply chain (from primary production to refining to transportation) must have sufficient capacity. This is not the case at any level, and the length of time required to build additional distribution infrastructure means it cannot be quickly remedied. In August 2004, world oil production equaled 83 million barrels per day, a level already well above that predicted by the model for that year. Achieving this supply level in 2004 required nearly 99 percent of the world's production capacity, making supply disruption in even one oil-producing area—or even the *fear* of supply disruption, such as that triggered by Osama Bin Laden's proclamation in December 2004 that he intended to attack the oil infrastructure of Saudi Arabia—a catalyst for rapid price increases.[6] In the 1970s, OPEC countries like

Saudi Arabia with ample spare capacity were able to operate as swing producers to help smooth out price spikes resulting from market turbulence. Today, Saudi Arabia and the other OPEC countries are producing full-out, and they can no longer act as shock absorbers for global oil supply. The oil shipping and refining infrastructure has been stretched progressively tighter over the years as well.[7] Gasoline prices in America rose sharply in 2004 and 2005 to their highest inflation-adjusted level since 1991, owing primarily to a shortage of domestic refining capacity.[8] The loss of refinery capacity in the Gulf of Mexico after hurricane Katrina in the fall of 2005 caused prices to spike to their highest levels ever in some areas and a return of gas lines throughout the American Southeast.

Similar if less acute conditions exist in the supply infrastructure of natural-gas and coal distribution. The real problem, however, is not just that fuel prices in 2004 and 2005 were 60 to 70 percent higher than their 2002 levels, a condition that might be explained by the expanding world economy over the interim period, but that they were substantially higher than those *at any time in the EIA forecast period through 2025*. Clearly, if prices for fossil fuels were to stay at or above the average levels of 2005, both world demand for and spending on energy would be heavily affected—and the EIA forecasts would be increasingly suspect.

The common response to this critique is that prices are cyclical: as prices go up and capacity is squeezed, energy companies are motivated to invest in refining and transport infrastructure. This argument has some validity, but in this case it overlooks two important points. First, the infrastructure is stretched thin because producers have been underinvesting in capital expenditures for nearly two decades. Fifteen years of low oil prices and some excess capacity in the early 1980s moved most oil producers to limit capital expenditures, a trend that prevailed in nearly every corner of the energy and electricity market. In fact, a recent IEA report calculated that to meet projected worldwide demand for energy, companies will have to increase their capital investment to over $500 billion annually between 2001 and 2030, a rate over twice that of the 1990s.[9] Today, reliance on the existing infrastructure gives producers very low levels of cost and depreciation on their capital base, reducing current costs of energy. The projected increases in capital spending to meet growing demand will add significant upward pressure to the cost of fossil fuels and the electricity they generate. Further, large new capital projects cannot be brought on quickly. It takes years to build new production infrastructure, sometimes decades, as in the case of hydropower projects, refineries, and nuclear plants. Supply disruptions and demand growth can have dramatic effects on prices in the meantime.

Second, there is the continuing problem of perverse incentives. Fossil fuels are a finite resource, and no amount of capital investment can change that fact. Additional investment will make little difference in the amount of oil or natural gas located under Saudi Arabia, Russia, or China. To maximize the value of these dwindling reserves, owners would like to get the highest price for every drop they sell, thereby reducing their incentive to use capital to lower prices through capacity expansion. Simply, lack of capacity increases prices. Furthermore, lack of capacity coupled with supply disruptions has an even more dramatic effect on prices—conditions well understood by the oil- and gas-producing nations of OPEC and Russia.

Looking at the assumptions underlying the EIA's predicted fossil-fuel prices through 2025 shows that key drivers of that energy-forecast model are already wrong. Even if prices fall back into the range predicted by the model, it is doubtful that the expected level of production will also be achieved. However, it is not very useful to say that the forecasts are wrong. What needs to be understood is why they are wrong and what other scenario is more likely to occur. Given the dynamics discussed previously, it is clear that fuel prices most probably will remain substantially higher than the EIA's forecasts. Given natural and infrastructure limitations on the production, transportation, and processing of fossil fuels, and given how vital energy is to all people everywhere, rising demand will drive higher prices. Three big questions arise: how far will prices rise, how quickly will they rise, and how volatile will prices be in the interim? None of these can be accurately forecast or adequately answered in this book, but the answers to these questions will dictate how severely energy consumers are affected by changing fossil fuel dynamics in the future.

Visions of a High-Price World

Higher energy prices, especially when coupled with volatile price changes, drive consumers to use less fossil fuel than they otherwise would have. This result is what the laws of supply and demand predict and what occurred in practice during the oil shocks of the 1970s. Higher fuel prices for one type of fuel (for example, oil) will motivate users to switch to fuels that are relatively less expensive. If prices for all three fossil fuels climb together and climb high enough, however, then the world will have to switch to renewable (or, as some industry observers argue, nuclear) sources, because these energy sources will have become relatively less expensive. In effect, higher and increasingly insecure fossil-fuel prices drive people to adopt cheaper and more reliable forms of local, renewable energy.

On the face of it, lower fossil-fuel usage and a faster push toward renewable energy may seem like a good thing, but such an adjustment is painful when it must be implemented rapidly—as demonstrated by the experience of energy shortages, gasoline lines, and high utility bills of the 1970s. In the short term, such supply and price shocks can wreak havoc on a society's income, growth, and prosperity. Unfortunately, given that *all* of the fossil fuels will be more expensive this time around, the next adjustment will be even more severe than the one experienced in the 1970s. The forthcoming changes will involve switching the 80 percent of the energy production infrastructure that relies on fossil-fuel sources, a change of quite another order.[10]

Large price spikes in basic energy prices will affect many individual and government priorities indirectly. In the face of high prices, corporate research and development budgets would be cannibalized to meet current consumption. It is also likely that people's concerns about the polluting effects of energy production will take a back seat to supplying cheaper power. Even today, when demand peaks or supply outages cause price spikes, such as during the California electricity crisis of 2000 and 2001, U.S. electricity makers receive special permission to restart highly polluting coal plants that normally remain turned off.[11] In the case of a sustained spike in the prices of fossil fuels, governments might well suspend or repeal environmental protections that restrict polluting generators. Already, the EIA forecasts an increase in annual global carbon emissions of 60 percent by 2025.[12] This forecast may prove too low, but at its projected level it is nearly double what is targeted in the 1997 Kyoto Protocol to the United Nations Framework Convention on Climate Change to reduce the worst impacts of global warming.

At the same time, many marginal consumers of energy, including vast numbers of people in the developing world, would be priced out of the market. Insidiously, the cost of nearly everything, including food and clean water, would increase for these nations because energy is the universal invisible ingredient. The cost of fertilizer, for example, is a key component in keeping most cultures from starvation and is tied to the cost of the natural gas used in its production and the diesel fuel used in its transportation. The economies of developing nations would therefore be severely strained by higher fuel prices. Either additional foreign aid would be required to maintain these economies, or they would find themselves in even deeper poverty. And competition over remaining global sources of fuels between the large developing economies of China and India, among others, and the industrial economies has the potential to create economic, if not military, conflicts.

In industrialized nations, the cost of electricity would increase. Fuel is some 15 percent of the cost of electricity, so for every 10 percent increase in the average cost of fuels, electricity prices are likely to increase by about 1.5 percent from fuel-price effects alone. Thus, a doubling of coal or natural gas costs from the baseline EIA forecasts would cause the projected price of electricity generated from that fuel to increase by 15 percent. In reality, the price of electricity would increase even more, since many of the nonfuel inputs to electricity production, from capital costs to transportation to materials, are also subject to changes in fossil-fuel prices.

Nor would industrialized nations suffer only from high prices for electricity and other commodities if fuel prices go up. Inflation in energy prices would result in broader inflation across the board and in a redirection of productive resources toward supplying basic energy, creating the dual effect of dampening economic growth and driving up the cost of borrowing for businesses and consumers. Not only would wealth be transferred from fossil-fuel-poor countries to fossil-fuel-rich countries, but the required focus on and rapid adoption of energy-saving mechanisms for homes and businesses and the forced retooling of energy generators to alternate fuel sources would dampen world economy and productivity.

Examining these forces suggests that the world's economy would sustain serious pressure from a sustained major rise in the price of any or all fossil fuels. No other conditions are necessary in this scenario, and even moderate price increases would have dramatic effects given the vital nature of energy. Even so, the economic comparisons that follow will use the EIA forecasts and assume that the price of energy and electricity in 2003 and 2004 are reasonable proxies for their future prices. Any variance will likely show this assumption to be overly conservative.

Electricity-Generation Economics

Some of the most common metrics for comparing across various electricity-generating technologies include installed cost per peak watt, cost of electricity generated, and cost of generation plus external costs, each of which is discussed below. Each of these methods of comparison is relevant or useful for a specific type of evaluation, and different users would rely on different measures to answer the questions of interest to each of them. Even when the appropriate methodology is chosen to understand a particular issue, the assumptions that feed into each of the different calculations must also be clearly examined to determine their relevance and impact on the results.

For example, one analytical method commonly used to evaluate the relative cost-effectiveness of various electricity generation solutions is to compare the cost of the generator required to produce a certain amount of peak capacity of electricity. *Peak capacity* is the maximum output that a given electricity source can produce at one time. Globally, some 4,000 GW of peak electricity-generation capacity exists today, and an average of 150 GW of new peak capacity has been added every year since 2000.[13] Comparing the costs of installing additional units of peak capacity for different technologies is one method of ranking their relative cost-effectiveness.

In addition to the cost of installing the electricity generator, the cost of operating the system over its life cycle can be significant and varies dramatically by technology. The two main components of operating cost are fuel and *maintenance*—a catchall term for all the nonfuel costs of running a plant. Broadly speaking, maintenance expenses include labor, overhead, repairs, and the periodic replacement of parts needed to maintain operations. Operating costs should also include the net cost of shutting down and dismantling a generator (and waste disposal when required) at the end of its useful life, but in practice this cost is sometimes overlooked or significantly underestimated. In addition, the lifetime cost of an electricity generator is also a function of purely fiscal variables, including the method and cost of financing the project and the assumptions made in discounting future costs and revenues for comparison to today's dollars. The analysis of the fiscal component of *operating cost* can be tricky because even small changes in financing and discount-rate assumptions can significantly change the projected economics of an electricity-generator installation.

By including appropriate installed peak-capacity costs, projected fuel costs, projected maintenance costs, and financing assumptions, it is possible to construct a measure of cost per kilowatt hour (usually measured in cents per kWh) of the electricity each produces—and this measure is often used today in the electricity industry for comparing across technologies. The range of generation costs per kWh for different technologies (including that of PV electricity developed in the previous chapter) are shown in figure 6.2.[14] It is this analysis that energy analysts typically refer to when suggesting that PV electricity is simply too expensive to compete economically in the global energy industry.

The cost estimates in figure 6.2 are subject to dispute because they include various assumptions about the operating environment over the long

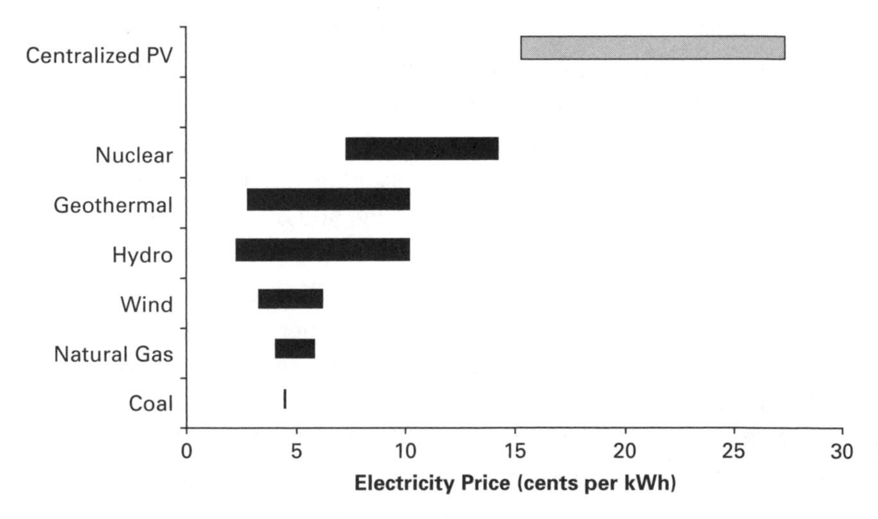

Figure 6.2
Utility electricity generation costs for various technologies (cents per kWh).

useful life of most electricity generators—predictions of future fuel and operating costs, guesses about methods of financing the construction of the generator, and so on. Without a comprehensive accounting of these embedded assumptions, it is difficult to judge the validity of any one cost-per-kWh estimate. Different estimates may even use different assumptions for the same variables, making cross comparison among the analyses prepared by different sources dubious unless all variables are well understood. Even when consistent assumptions are used, many of these types of electricity generation, particularly those reliant on fossil fuels, retain a substantial risk of future changes in the cost of the fuels that they need to operate throughout their useful lives.

The real limitation to the cost-of-generation method comparison is that it assumes as a constant the cost of getting the electricity from where it is generated to where it is used—which can be up to half of the cost of electricity to the end user. It assumes that this end-user cost is irrelevant to the decision of which type of electricity to produce and that generation cost is the only relevant variable. In the past, analysis done only from the point of view of the utility was reasonable. Choosing what kind of generator to install could pretty safely assume that, regardless of which was selected, the power would have to be delivered by the same method—a modern electricity grid.

Cost per kWh generated has therefore been a standard measure of comparative cost for many years by electric utility companies, but industry fundamentals are changing as the potential to shift location of electricity generation increases. Today, it is increasingly obvious that distributed generation technologies (such as small reciprocating engines, mini wind generators, and photovoltaics) are rational options for electricity consumers that may not require delivery through the grid. As a result, the old cost-of-generation metric does not work reliably anymore for an end user or even for a utility-adding generation capacity on the customer's side of the grid when comparing potential forms of electricity generation.

A new and more useful metric that measures cost per kWh *as delivered to the end user* can be constructed instead by adding transport cost to the standard calculation of cost per kWh of centralized generation. The cost of grid-based, centralized generation can then be directly compared to the cost of independent, on-site distributed generation on the same basis.

As would be expected, on-site distributed electricity is significantly more cost competitive with conventional electricity under this type of comparison than it would be under a utility's standard cost-per-kWh-generated method. In practice, the cost-per-kWh-*delivered* method of comparison allows for an energy consumer that is evaluating the purchase of distributed energy technologies such as PV to compare the cost of the PV-generated electricity to the cost of the grid electricity it replaces. Evaluating electricity economics from the point of view of an end user when a distributed source of electricity generation such as PV is available is discussed in the next chapter.

Yet another method of comparing the costs of electricity sources includes the *external costs* of the generation technologies. External costs are the quantifiable societal or environmental costs of each electricity source, including the direct costs of pollution (property damage, health-care costs, deaths, and the like), other environmental damage (for example, destruction of land for strip mining, transmission lines, and reservoirs), security costs for protecting fuel supplies and nuclear facilities, and the costs of disruptions in electric supply for centralized generation. These external costs are often cash costs, though many are paid for by taxes and government borrowing and not directly by the energy consumer. If these external costs were included in the costs of various forms of electricity generation, the picture of the relative costs of different technologies and their relative attractiveness would change, increasing the cost of fossil fuels by between 30 and 90 percent for natural gas and 55 to 400 percent for coal.[15]

Table 6.1 shows the various methods of comparing sources of electricity generation and when each economic comparison would be appropriate. Each of these comparison methods—cost per peak kW installed, cost per kWh generated, cost per kWh delivered, and cost of generation plus external costs—is useful for a specific type of analysis. Typically, comparisons within a technology—standard coal-burning versus fluidized bed coal-burning, for example, or comparisons among various PV modules—would use cost per peak kW installed to see which choice is better. Cost per kWh generated is more useful when comparing various generation sources for a utility or industrial off-grid customer where distribution is a fixed cost. Cost per kWh delivered is the best metric for an end user who is considering installing generation capacity to supplement

Table 6.1
Survey of economic assessment methods to compare electricity costs.

Comparison Method	Type of Analysis	Point of View	Example
Cost per peak kW installed	Comparisons within a generation technology class	Any purchaser of a predetermined type of generator	Standard coal burning versus fluidized bed coal burning; thin-film versus silicon PV
Cost per kWh generated	Comparing various energy sources for electricity in which distribution is a fixed cost	Utility or industrial off-grid customer	Choosing between a new nuclear power plant or geothermal plant to provide base load electricity
Cost per kWh delivered	Considering installing generation capacity to supplement or replace grid electricity	End user (home, business, or factory owner)	Comparing electricity from a rooftop PV system versus equivalent purchases of grid electricity
Cost of generation plus external costs	Comparing the environmental and social impacts of various energy choices	Governments, policy analysts	Evaluation of the total environmental impacts of fossil fuels versus renewables

or replace grid electricity—for example, a homeowner weighing a rooftop solar system against equivalent purchases of grid electricity. And external costs are the type of analysis that governments and policy analysts should use to account for the social and environmental effects of various energy choices.

The Cost of Time

Returning for now to the type of cost-of-generation method predominantly used by utilities for centralized electricity generation, comparison among various energy technologies must also distinguish how and when the electricity provided by each of these generation sources is used.

Some centralized generators such as base-load coal and nuclear-power plants run nearly all the time, in many cases over seven thousand hours per year, shutting down only for required testing and maintenance. Others—such as some natural-gas plants and some hydropower projects that can be run for less than fifteen hundred hours per year, in some cases—are affordably run only at times of high demand. Still others, such as wind turbines, are inherently intermittent, and site conditions will determine how many hours per year they can be used to generate electricity.

The relative cost-effectiveness of each electricity-generation technology thus depends on the type of power flow each individual generator provides and how many hours per year it is operating at its peak capacity. The ranges for the cost of electricity shown in figure 6.2 assume that the generator is being used at its optimal output. In the world today, the average electricity generator is used only about half of its potential capacity, with nuclear and coal plants used substantially more than half the hours in a year, and natural-gas generators used substantially less than half of them. Economically speaking, these part-time generators are not run optimally, and therefore their electricity is correspondingly more expensive because the fixed cost of purchasing and maintaining the equipment is spread over fewer useful hours. Comparing a technology, such as PV, that is used only a portion of the day with a generator that is assumed to be working all of the time is incorrect.

Another common misperception using the analytic methods above is that a generation method is economic only when it is as cheap as the cheapest types of electricity generated by any other form of energy. This is not the case. An appropriate comparison of the cost-effectiveness of any form of energy would compare the cost of that energy with the specific type of load it replaces—base-, intermediate-, or peak-load.

PV electricity is generated during the times of day when the sun is available and in general proportion to the level of sunlight. As previously discussed, this solar availability is highly correlated with the electricity-demand patterns in a typical modern electricity system. More electricity is used when people are awake and productive, and this occurs most typically during the middle of the day. Figure 6.3 shows this typical demand pattern throughout the day and shows how different types of electricity-generation methods are used to meet different portions of the

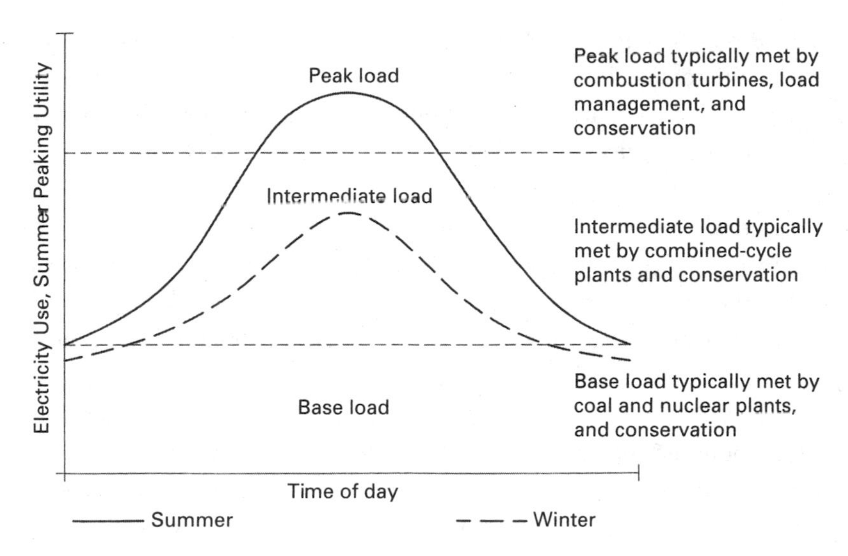

Figure 6.3
Typical load curve for an electric utility.
Source: Wisconsin Public Service Commission.

demand. The cost of generating electricity rises in proportion to the level of demand, and the wholesale market for generating this power sees significant increases during the middle of the day, precisely when the sun availability is at its highest.

Thinking through the economics of electricity generation helps clarify why certain generators are used as different times and for different types of load. Among the various electricity-generation technologies, natural-gas generators have high operating costs compared to their initial capital cost, with much of their expense in the fuel used for generation. For this reason, natural-gas generators are used more frequently to supply variable power needs as intermediate- or peak-load generators. Coal and nuclear plants, on the other hand, have large capital costs, fairly high operating costs, and comparatively low fuel costs. Owing to their cost characteristics, coal and nuclear plants are more economically used as base-load generators.

Therefore, comparing the cost of PV to constantly running coal and hydropower base-load generators is not a meaningful comparison for the way PV generates electricity and is used today. From a utility's perspective, the type of electricity that PV replaces is the intermediate- and peak-load power that is generated using power plants that are being run only a portion of the time and that therefore produce more expensive electricity. These intermediate- and peak-load generators produce electricity at a wide range of prices depending on the number of hours per year they are used, the type and cost of fuel, and the age of the equipment. But some intermediate- and peak-load generators used by utilities today are producing electricity more expensively than a comparable PV system would in the same location and size.

Continuing to look at the economics of generating energy from a utility's point of view, base-load power can be generated under optimal utilization of the generators at the costs in figure 6.2, which represent on average about 65 percent of the total electricity demand in modern industrial economies.[16] However, utilities must provide the more expensive intermediate-load power using part-time natural-gas generators, usually during daylight hours, which represents some 30 percent of all of the electricity supply.[17] It is precisely this type of expensive, daytime intermediate-load power that solar energy and photovoltaics by their nature replace and should therefore be economically compared to. Figure 6.4 shows how the cost of PV compares with both base-load and intermediate-load electricity.[18]

It is not necessarily the case that utilities will choose to use nonconcentrating PV technology as their first choice of centralized solar-electricity generation for daytime electricity. Alternative solar generation technologies are available and cost-effective at the industrial scale. Large concentrating PV applications and solar thermal generators, such as the Luz SEGS plants in southern California, are viable options as well. In 2005, a number of new utility-scale solar thermal plants were announced in California, and over 130 MW of additional solar thermal electricity is expected to come on-line by 2007.[19] Plans for up to twenty MW of new concentrating PV plants have been announced in Spain and Australia, as well as the American Southwest. Each of these technologies should enjoy additional interest in the coming years as utilities increasingly look for solar alternatives to provide their intermediate-load.

With various solar-energy technologies currently providing less than 0.05 percent of total electricity generated globally, the economic potential for utilities to deploy them to supply industrial intermediate-load and peak-load shaving will take decades to meet. As PV slides down its expe-

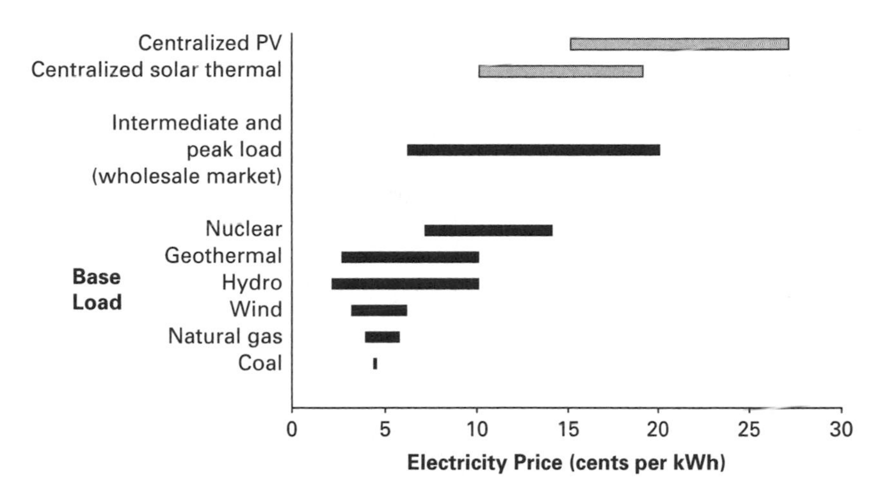

Figure 6.4
Utility electricity generation costs (revisited), including relevant wholesale power prices and solar thermal electricity.

rience curve to become relatively cheaper than the fuels that utilities currently rely on to generate intermediate-load electricity—primarily natural gas—utilities will increasingly look to solar energy and PV as an energy solution.

Projecting Utility Economics

Projecting how the economic relationship between centralized base-load generation, intermediate-load generation, and the cost of PV systems will evolve over time shows an interesting transition over the next couple of decades. Figure 6.5 shows a forecast by Winfried Hoffmann, former head of the European Photovoltaic Industry Association, which estimates these changing economics. Under these projections, PV in the sunnier locations of Europe (the 1,800-hours-of-sun-per-year curve on the graph) is already beginning to provide cost-effective alternatives to intermediate- and peak-load power (which the figure refers to as *utility peak*). Within fifteen years, PV will be cost-effective nearly everywhere for this type of electricity even though PV will still be providing only a fraction of that potential. Within twenty-five years, PV will become cost-effective in base-load generation (which the figure terms *bulk cost*), opening up additional market potential for this technology.

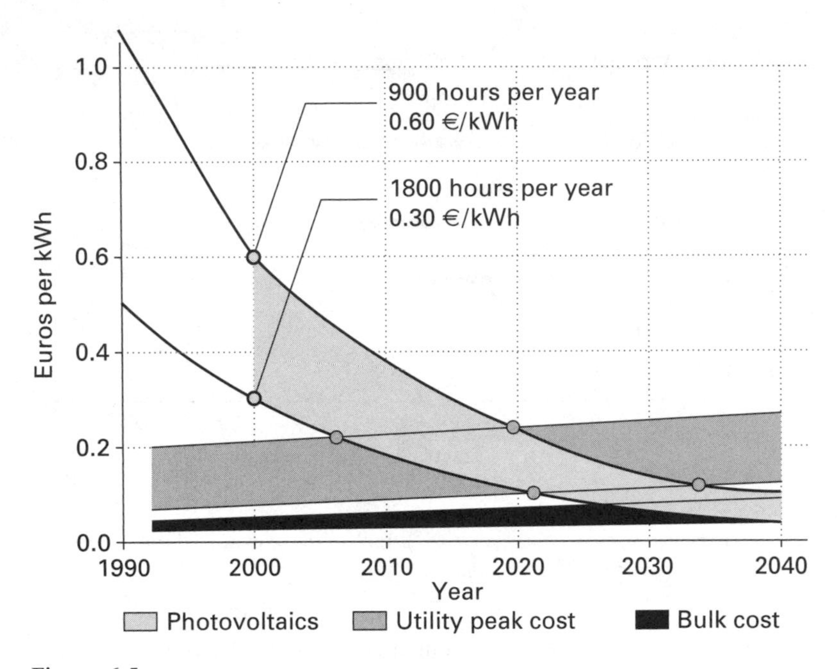

Figure 6.5
Projection of PV competitiveness in Europe through 2040.
Source: Winfried Hoffmann.

Some utilities will realize competitive economics in solar energy before others, based in part on the amount of available sun and land to site the systems. Economically, the first systems of this type will be large-scale installations in desert regions with intense, reliable sun. Tucson Electric Power is an example of such an early adopter, feeding 4.6 MW (peak) into the grid from a photovoltaic installation in Springerville, Arizona, that is one of the world's largest PV generators.[20]

Storage Eventualities

The combination of all the solar technologies—centralized PV, solar thermal electricity, and concentrating solar power (as well as distributed PV for end users, which is discussed more fully in the next chapter)—will inevitably grow to supply larger proportions of the grid's electricity. However, there are potential technical limits to widespread adoption of intermittent sources of electricity beyond 15 percent of total grid capacity without the added inclusion of energy storage solutions to smooth out these intermittency issues.[21] In the meantime, utilities that build centralized solar-electricity-generating plants can manage their electricity load in the same way individual grid-connected consumers can—that is, sell their daytime output to the bulk power market and repurchase cheaper power at night. If utilities eventually install a larger proportion of intermittent electricity generators, they will have to adopt large-scale energy-storage applications—pumped hydro, compressed air, hydrogen fuel cells, or advanced flywheels—to supply power during periods of low or no sun.

Yukinori Kuwano, president of Sanyo Electric, has suggested an intriguing nonstorage alternative called Project Genesis. This global energy-infrastructure project would interconnect the whole world's electric grids through superefficient, high-capacity, intercontinental transmission lines.[22] Genesis would be the logical extension of the system integration that has been going on since the early days of the electricity industry and would allow PV systems to supply a large amount of the world's power through the existing grid infrastructure. The day side of the world could sell power to the night side of the world, reducing the amount of additional electricity storage needed to smooth solar-availability intermittency. The proposal has potential problems—for example, Genesis's superlines would have to be added to the Energy Information Administration's list of vulnerable energy "chokepoints"—but highlights that there are both opportunities and need for innovative thinking in the evolution of the electricity-generation infrastructure.

From a utility's point of view, the economics of centralized electricity generation are beginning to change. Never in history have utilities had the option of economically generating the most expensive, daytime intermediate and peak-load portion of their electricity demand with a renewable technology that is specifically available during the times when that load is demanded. The expectation of improving economics of solar electricity, for both PV and thermal applications, means that utilities will increasingly enjoy advantages in deploying solar electricity over the next few decades, even if the cost of conventional electricity-generation methods do not substantially increase.

However, the improving cost of industrial-scale, centralized solar energy and the potential for increased costs in conventional electricity are not the only forces that utilities will face in the coming decades. The economic characteristics of small-scale photovoltaics could begin to unravel the economies of scale that Edison's electricity transmission created over the last century, with dramatic effects on utilities. The next chapter describes the emergence of a new distributed-electricity economics that will revolutionize the energy and electricity industries worldwide.

7

The Emergence of Distributed Economics

Since the beginning of the twentieth century, electric utilities and the economies of scale they possess to cost-effectively provide energy in the form of electricity to customers has been a strong growth catalyst for industrial economies. Today, a new generation of technologies, predominantly grid-connected PV systems on the homes or buildings of end users of electricity, provide an alternative source of vital electricity and a new economic comparison to evaluate their competitiveness. An individual or business that is evaluating whether to switch to a distributed-energy technology (such as on-site PV) need only compare the relative costs of two basic options—to stay with centrally generated utility power or to adopt a PV alternative.

As experience is gained and the scale of production for the emerging technology of distributed PV increases, the cost per kWh delivered of solar energy has dropped and will continue to drop as a function of cumulative global production. As the forecasts of chapter 5 show, layering in the expected market growth rate provides an approximation of when certain cost levels will be reached. These times could move forward or back by a couple of years, but the fundamental trends will continue: within a decade to a decade and a half, distributed PV will be the cheapest electricity option for a majority of residential electricity consumers in the world.

Because each potential grid-tied market has slightly different characteristics, including system and installation costs, discount rates for financing the systems, and the amount of sun available in each location, it is possible to determine where and when customers in particular markets will begin to find it cost-effective to adopt PV. The grid-tied market is a mosaic rather than a monolith—that is, many markets rather than one. Despite this heterogeneity, only a few key factors—PV system cost, *insolation* (average sunlight), and the local price of grid-based electricity—determine which markets will become cost-effective for solar energy early on. These three factors define a market hierarchy in which customers of markets near the top will find it cost-effective to switch to solar technologies sooner. This hierarchy shows where early adoption can be expected and how solar energy will unroll from those initial markets.

Calculating the Cost of Distributed PV

Based on the cost-of-delivered-electricity comparison discussed in chapter 6, grid-tied PV electricity becomes a cost-effective technology when its cost on a customer's site, either home or building, drops below that of local grid electricity. This scenario has become the case in much of Japan today owing to impressive growth in the market, the resulting drop in PV system prices, and some of the most expensive grid electricity in the world. Many new residential PV systems in Japan are providing electricity to their owners at or below the cost of grid electricity without any subsidy or support from the government. As a result, a large and important market for distributed PV has been established with 46 million households at or near cost-competitiveness, and for that reason the rapid growth of the Japanese PV markets is expected to continue.[1]

It is not obvious that PV electricity's cost competitiveness in Japan means that PV technology can meaningfully compete in the larger world-energy industry. For starters, the Japanese residential market has some of the highest prices of grid electricity in the world—an average of twenty-one cents per kWh.[2] With most countries in the OECD selling electricity at less than half that price, PV would appear to have a long way to go to become cost-competitive in global markets. In addition, Japan is not a sun-rich country compared to many other industrial or developing nations, and PV may find better economics in other sunnier locations. To understand the relative competitiveness of PV across various locations requires understanding exactly what determines the cost of PV electricity and correcting a few of the common myths regarding its cost structure.

More so than any other form of electricity generation, PV's natural economic disadvantage is that nearly all of its costs are incurred in the installation of the modules and components. Without any fuel costs and almost negligible maintenance costs owing to the solid-state nature of PV panels and inverters, over 90 percent of the lifetime cost of a system is paid up-front. To impute the price of PV-generated electricity requires knowing three variables about a given installation—installed system cost, how the system is financed, and the amount of sun available at the location to be served. Because each of these three factors is known with relative certainty prior to an installation, one of the underappreciated advantages of PV electricity technology is that it provides a reliable estimate of lifetime electricity costs.

Today, the cheapest price for an installed grid-tied system ranges from under $6 per watt for Japan and Germany to as high as $7 per watt in the United States.[3] The cost of the PV panels and the necessary supplemental hardware (such as wiring and inverters) are relatively, though not exactly, consistent across locations. However, the cost of installing the system can vary widely based on the competition and cumulative experience in the local market.

Financing the purchase of PV systems can vary both in the term and the interest rate available, and many estimates use incorrect assumptions for these variables. One common error made in many calculations of the true cost of PV electricity is using *nominal* interest rates instead of *real* interest-rate figures. Real interest rates are the inflation-adjusted cost of capital, not the nominal rate written into loan contracts or mortgages. Real interest rates are calculated as the nominal rate of the financing less expected annual inflation in prices of goods and services over the life of the loan and almost always result in a lower cost of capital than using the stated nominal rate. An economic calculation that incorrectly uses nominal rates to determine economic viability actually builds in a declining real cost of electricity over the life of a PV system and therefore overstates the current cost of PV electricity. Depending on other variables, this correction alone can bring down the imputed cost for solar energy by 20 to 25 percent over many of the commonly quoted prices.

Calculating the real annual cost of a particular PV system and dividing by the average annual sun in a given location gives a cost of electricity on a cost-per-kWh basis, which can range from twenty-one cents in Japan to around sixteen to nineteen cents in the sunnier American Southwest for an unsubsidized PV system. Using these numbers overstates the price of PV because installations in many markets today enjoy system subsidies or incentives and in the United States the deductibility of home mortgage payments, which reduces the actual after-tax cost of the system for customers.

Using these calculations, the experience-curve tool discussed in chapter 5 can then be applied to determine the future expected cost of distributed PV in various locations based on their level of solar resources and the type of financing used to pay for PV systems. Specific calculations for the economic comparisons in various markets are developed later in this chapter after an examination of the current and future prices for the grid electricity that PV competes with as a source of distributed electricity generation.

Distributed PV versus Grid Electricity

Electricity rates in industrialized OECD nations decreased on average between 1994 and 2000 from an average of 11.6 cents per kWh to an average of 10.5 cents per kWh, a trend related to a decline in fossil-fuel prices over that period and the concurrent move toward utility deregulation.[4] Given this decline in utility electricity rates, it is reasonable to ask whether experience curves might apply to the grid as well and how this might alter the earlier analyses and forecasts. The answer is that experience-curve analysis is not an appropriate forecasting tool for the bulk of the conventional electricity generators that make up the modern electricity infrastructure, including fossil-fuel, nuclear, and hydropower generators. The cost of conventional electricity is primarily a function of the cost of fuel and capital inputs rather than the historic volume of production; most of these have already captured their optimal economies of scale. Experience curves should be used to forecast future costs of young technologies as they achieve initial market acceptance, but today the grid is well established and market dominant. Therefore, it is more appropriate and useful to forecast future electricity rates of conventional generators using the projected costs of the inputs in their production—fuels, maintenance, and capital investments.

As mentioned earlier, both fuel prices and the capital requirements to maintain and upgrade the current electricity-generation infrastructure are likely to increase. Indeed, they are already increasing and affecting the rates that utility customers pay. In America, for example, electricity rates as of mid-2005 were 7 percent higher than 2003, and similar changes are occurring throughout much of the industrialized world.[5] Many U.S. state-level rate caps dating to the late 1990s are beginning to expire, and many regulated utilities are planning substantial rate hikes, particularly for residential customers. This price trend suggests that it is extremely unlikely that grid-electricity rates will drop below current prices in the foreseeable future. In fact, they are more likely to rise. For the purpose of the economic comparisons throughout the rest of this chapter and in line with a conservative forecasting approach, today's electricity rates are assumed to be a reasonable proxy for future rates. This may be an overly conservative assumption, and in the event that utility rates rise substantially or quickly, distributed solar energy will become cost-competitive with grid electricity sooner.

If the cost of established technologies such as nuclear and fossil-fuel electricity is assumed to be flat, then what about the cost of the nonsolar forms of renewable energy that are increasingly deployed in the current infrastructure at decreasing cost? Several technologies, such as wind and geothermal, are tracing dramatic experience curves as they grow in volume and market adoption. Just as for PV, learning in these technologies through increased scale and sophistication will continue to move these technologies toward (or beyond) cost-competitiveness with existing generation methods even on an unsubsidized cash basis. Though wind and geothermal continue to get cheaper, their ability to compete with distributed solar energy is limited by the method of their use. To be cost-effective, these new renewable solutions of wind and geothermal energy must be deployed on a large scale and as part of the larger existing utility infrastructure (so that they provide their electricity to end users through the electricity grid). Obliged to grow inside the installed base of capacity and transmitted over the same electricity grid, these solutions can be expected to have only a small impact on the total cost structure and therefore on the rate that a utility charges for electricity. This conclusion can best be clarified via the economic concepts of marginal cost and average cost.

Marginal cost is the cost of the next unit of something to be purchased. As applied in the power industry, it refers to the cost of each additional electricity generator at a given point in time that a utility or energy user installs. Cost per kWh generated, described in the previous chapter, is the standard utility yardstick for measuring the marginal cost of generation capacity. As new renewable solutions slide down their experience curves and become more cost-effective, utilities should be able to add generation capacity in the form of wind and other renewables at marginal costs on a cost-of-generation basis that are competitive with those of the existing nuclear or fossil-fuel generators.

In contrast, *average cost* is the average cost of all units produced—in this case, the entire portfolio of electricity generators that a utility or user owns. Declining costs of centrally generated renewables (like wind, hydro, biomass, and even the centrally generated solar electricity methods discussed in the previous chapter) do not materially affect projected utility rates for electricity because utilities generally cannot charge their customers for electricity at a rate proportional to the *marginal* cost of energy production. Instead, utilities typically charge their customers a fixed rate for electricity and must charge at the *average* cost of their entire portfolio of generators plus profit.

From another perspective, since the average age of the grid's components is fifty to sixty years, the replacement rate of these assets not including growth should be the inverse, about 2 percent per year.[6] Even if a utility bought 100 percent of its replacement generating capacity in the form of declining cost renewables for ten years—a feat not even close to being achieved in any country—after ten years the percentage of the utility's proportion of renewable electricity could grow by only 20 percent in that time. Eighty percent of the grid would remain under the old cost structure, which is primarily dependent on fossil fuels and installed capital costs. Despite a utility's declining costs for centrally generated new renewable sources of electricity and the understandable optimism of many in those industries, these new renewable technologies will not reduce the average cost of electricity much (if at all) in the next couple of decades. However, it is still useful to develop and deploy these generation technologies wherever they are economic to do so versus competitive fossil-fuel generators because their deployment will mitigate potential increases in the cost of fossil fuels as well as generate electricity with dramatically reduced pollution.

Photovoltaics are different. Distributed applications, for which PV is nearly as easily deployed as for centralized applications and in many ways cheaper, are among the few applications that can be installed and run at marginal cost for the end user without being subsequently dragged down into an average cost for the utilities' portfolio of generators. Homeowners and businesses can install PV on-site, circumventing the entire existing supply chain of electricity generation and distribution. Each kWh that the PV system generates replaces each kWh of grid electricity on a one-for-one basis, allowing the full marginal-cost savings of the system to be captured. As the cost of PV systems continues to decline, they will be able to compete with grid electricity on the basis of marginal cost, immune to utility-scale average-cost considerations—a powerful economic driver for the future of distributed electricity generation and the entire energy industry. By comparing the forecasted marginal cost of distributed PV electricity and the current and expected average cost of centralized generation by utilities, energy policy makers can begin to get a sense of future energy-industry dynamics. With a grasp of these dynamics, they can predict how the photovoltaic market will likely unfold in the coming decades.

Mapping PV's Future

Each specific geographic area where solar power might be adopted reveals different obstacles and opportunities. A few key considerations determine a natural hierarchy of places where PV adoption should initially occur in the absence of government support or subsidy programs. Beyond the installed cost of the PV system, two other issues determine the cost-effectiveness of PV installations in a given location—the amount of sun at that location (*insolation*) and the cost of conventional grid electricity at the specific site to be served. These three factors—system cost, insolation, and the local price of electricity—are independent of each other and must be considered to understand how solar energy will progress in a given market.

The first main consideration of economics for any PV system is cost. Nobody says that solar energy will ever be "too cheap to meter," as was famously said of nuclear power in the 1950s, but as the experience curves of chapter 5 describe, hardware costs are coming down steadily thanks to technological learning and growing economies of scale. The real cost of a solar PV system is what a buyer pays after adding up all monetary costs (hardware, installation, and so on) and subtracting government rebates and tax breaks, if any. While important in determining system economics, hardware cost is a minor factor in trying to predict in which geographic markets PV will become cost-effective first because it is not strongly related to the place in which the system is installed—at least within the industrial world. If PV cells sell for a certain price in Japan or Germany, they will sell for a similar (though not exactly the same) price in Arizona or Madrid. Similar to digital cameras or other commoditized consumer goods, solar panels and remaining system components such as inverters and wiring can be made economically in many locations. As a result, their production ultimately will move to where slight cost advantages can be gained via cheap labor, cheap capital, and cheap energy.

As for incentives, different governments offer various options to promote renewable energy, which are discussed in detail in chapter 9. Financial incentives have already made it cost-effective to install PV in Germany and Japan, enabling these countries to become leaders in the current industrial shifts. In the United States, incentives vary widely by state, with some offering hefty cash rebates on solar PV systems—for example, $4,500 per kW of peak solar capacity installed in Los Angeles and $5,500 in New Jersey, which represent a range of 50 to 70 percent of system cost.[7] On the other hand, twelve states as of 2004 did not even have useful net-metering laws allowing small, grid-connected systems to link with local utilities.[8] Areas with better incentives will see more and faster growth than those with poor incentives or with regulatory roadblocks, but subsidies are tricky to forecast. New incentives can suddenly improve a location's attractiveness for PV system installations, but existing programs can expire, quickly dampening motivation. Subsidies in Germany and many U.S. states seem stable, but Danish government support for all types of renewable energy was essentially abolished after a change in government in 2001.[9]

The amount of sun that a location receives is a major factor in determining the cost-effectiveness of a PV system. Since cost of installation is similar across all locations in the industrialized world, if one particular site gets twice as much sun as another, then the first site should generate approximately twice as much electricity. Insolation is measured in kWh per square meter for a fixed time period (usually a day or year) and is a function of latitude, climatic conditions, and (over short intervals) time of year. Knowing these three variables allows for a rough estimate of how much electricity a PV system at any given site can generate. Using computer models, maps can be generated that show bands of insolation over large areas. Figure 7.1 shows a U.S. insolation map. Such maps can be used to make rough estimates, but actual insolation measurements taken in a specific location are invaluable for predicting the output of any particular PV system.

For end users of distributed PV electricity, economic comparisons also require knowing the local grid electricity rates, which vary widely among even the industrialized nations according to energy-source availability, location, and installed infrastructure. These rates range from six cents per kWh in South Africa to thirty cents in Denmark, creating a wide range of market opportunities for solar energy.[10] Within the United States, residential utility rates range from 6.2 cents per kWh in West Virginia to 19.5 cents in Hawaii.[11] The range of these residential utility rates and the cost for PV systems at today's price are shown in figure 7.2.[12] This figure also shows the range of costs for distributed PV electricity generated at today's best price of $6 per peak watt and potential future cost of PV of $3 and $1.50 per peak watt installed, respectively,

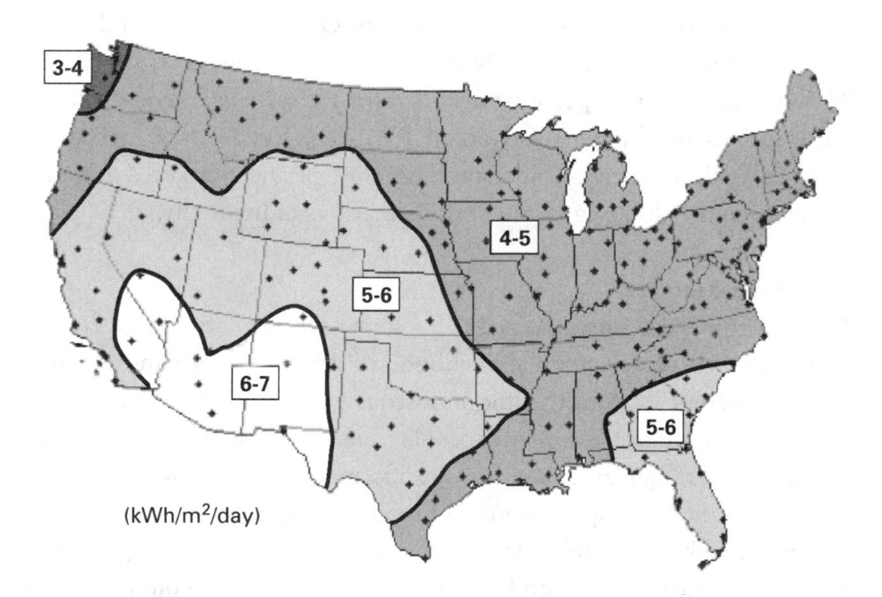

Figure 7.1
U.S. insolation map (annual)
Source: NREL (2003).

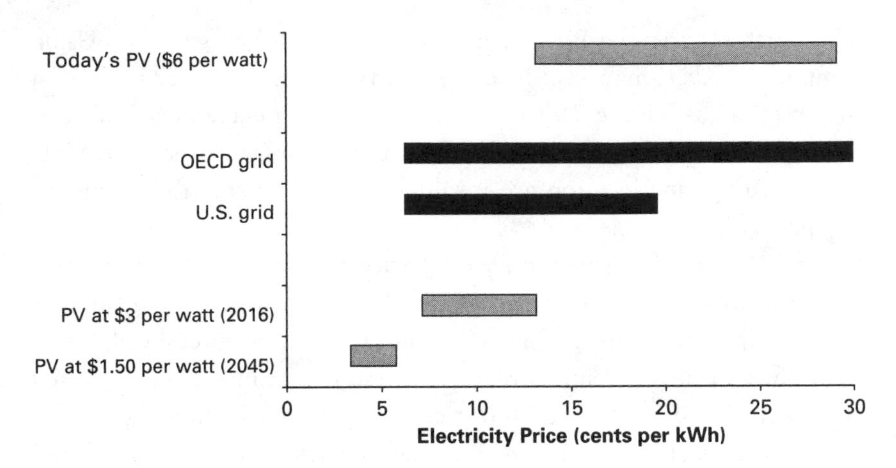

Figure 7.2
Range of grid electricity prices in OECD nations and U.S. states compared to end-user PV electricity costs at today's best price of $6 per watt and forecast future prices of $3 per watt and $1.50 per watt.

levels that the projections in chapter 5 would predict to occur by the years 2016 and 2045, respectively.

This comparison shows that a number of OECD countries and states in the United States can already cost-effectively adopt distributed PV. Within a decade, at $3 per watt installed for PV systems, the majority of the locations would find distributed PV cost-effective even at today's price of grid electricity. Understanding the specific market dynamics, though, requires looking at the various markets individually.

The Major Markets

Three factors—real unsubsidized PV system cost, insolation, and cost of grid electricity—determine the likelihood of market growth and maturation in different locations in the industrial world, although there is no precise equation relating them to solar market growth. These factors provide only rough guidance for specific market transitions. All three factors need not line up for a market to grow: any combination of them may be enough given their relative strength. In fact, the two largest markets for PV today, Japan and Germany, are among the countries least likely to find solar power economic if simply the amount of average sunlight is considered. Yet their alternatives are severely limited (Japan imported 81 percent of its primary energy in 2004, Germany 61 percent), and their governments have made bold strides to compensate.[13] Directed incentives and the high cost of grid electricity have created strong solar demand in both countries. On the other hand, with a combination of high sun, high electricity costs, and state incentives to pay for half or more of system cost, California leads the development of the American PV industry. Some Japanese manufacturers are recognizing this market potential, and Kyocera recently completed a maquiladora plant in Tijuana, Mexico, to supply solar PV cells to the growing California market. Hawaii is a natural market for solar energy because its tropical location provides good insolation and, as an island, it has the most expensive grid electricity in America.

In Europe, the Mediterranean countries are likely candidates for PV adoption because of their large insolation and scant domestic fuel resources. Some Spanish municipalities, such as Barcelona, are installing large grid-tied PV systems on the roofs of public buildings. With electricity rates nearly as expensive as Germany but with much more sun, Spain will quickly begin to realize the economic advantages of PV. Italy has high electricity prices too but has done little in terms of providing robust financial incentives to stimulate PV demand. Greece already enjoys electricity that is cheap by OECD standards, but its isolating geography and hundreds of habitable islands provide interesting opportunities for local PV applications—particularly given the amount of consistent sunlight available. Conversely, the Scandinavian countries have ready access to both fossil fuels and hydroelectricity that keep their power costs low. While they get a fair amount of sun, it comes disproportionately in the summer months, limiting PV's effectiveness unless large storage solutions are employed.

In America, the next candidates for locations where solar energy will grow after California include the remaining high-sun states of the Southwest (such as Arizona, New Mexico, and Texas) and the expensive-electricity states of the Northeast (such as New York, New Jersey, and most of New England). Several of these states also have strong state-level incentives for PV adoption. Many Southeastern states that have high levels of sun, particularly in the power-heavy summer months, but more moderate electricity prices should begin to see sustained growth in PV in coming years depending on individual state policies. The places least likely to go solar in the United States in the near term include the Midwest and Pacific Northwest because these states have average insolation and below-average electricity prices. Strong government support such as Washington State's recent adoption of a feed-in tariff for renewable energy, however, could shift the balance sooner than expected.[14]

Predicting Market Crossover

To get a sense of when applications in different locations and industries will become economically competitive, it is helpful to forecast longer-term experience curves and market growth for photovoltaics. Using the forecasts of the cost of distributed PV electricity through 2040 in chapter 5, figure 7.3 shows dropping prices for PV across various U.S. cities with different combinations of insolation and grid-electricity rates.[15] This figure shows the combination of various sun and electricity rates that would be economic at a different PV system cost levels, also called *isocost*

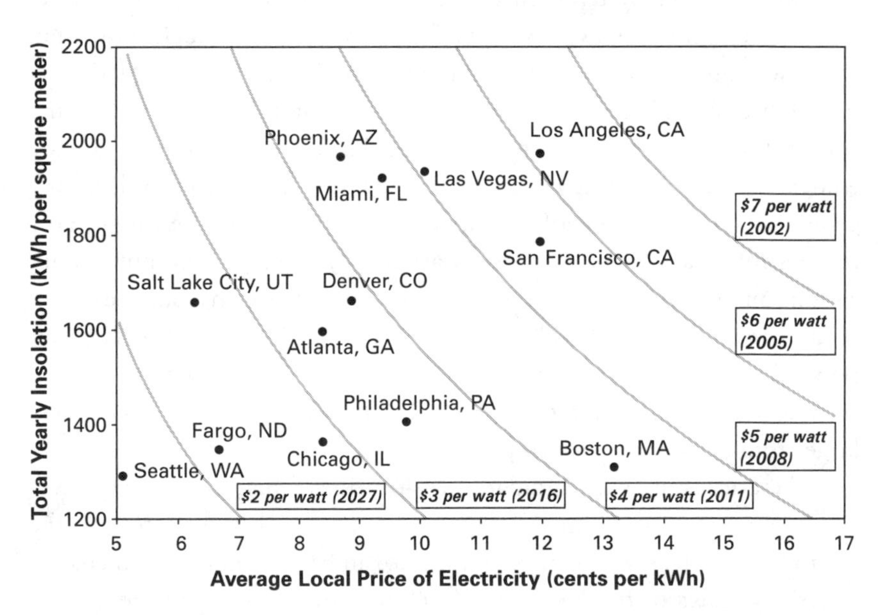

Figure 7.3
PV isocost curves for the United States at various cost levels of PV systems and how PV economics for various cities will change over time.
Source: EIA (2005); NASA.

curves by economists. Residents of any city that is to the upper and right of a curve would find PV an economic choice at that curve's price level. For example, residents of Los Angeles would find PV cost-effective at $6 per watt, but residents of Las Vegas would find PV cost-effective at only $5 per watt.

It is instructive to see how quickly many cities and millions of customers begin to find PV a cost-effective solution, even without the benefit of government subsidy programs. Assuming these cost reductions are right, within a decade, residents of Atlanta, Denver, Miami, and Philadelphia will be able to replace grid use with on-site PV. By 2016, Los Angeles will enjoy cost-effective PV at half the cost of grid electricity. Over the next fifteen years, the number of locations and customers that can economically shift to PV will accelerate much faster than the U.S. PV industry can scale up production, which will provide a strong growth incentive to the domestic PV industry.

Industrial Users

The final energy and electricity users—beyond utilities and distributed end users—are the industrial customers that provide their own local sources of electricity generation and that comprise as much as 10 percent of the electricity supply in industrialized countries. These customers will be the last to find the adoption of local PV economically compelling owing to the low cost of today's industrial-grade power, which can be as low as a few cents per kWh in some locations. Eventually, depending on the rate of decline in PV prices and the increases in the cost of traditional fossil fuels, PV can supply these users when coupled with adequate energy-storage technologies such as hydrogen fuel cells. Even before reaching economic equivalence, industrial PV has some added advantages that may accelerate adoption by these customers. Once PV has been installed, the cost of the electricity PV produces is fixed. This characteristic should have value to energy-intensive industries that view volatile or rising fuel prices as a major business hazard.

Energy-intensive industries also have the option to relocate to geographic areas where the richest solar resources decrease the cost of generating PV electricity. This strategy is similar to current practice where companies colocate near a primary cheap energy source based on today's energy choices. For example, Finnish paper mills site near cheap hydroelectric power, and Pennsylvania steel mills often locate near coal mines. As the cost curves of industrial-scale PV and conventional sources of energy converge, such factors will increasingly weigh in on siting decisions.

An additional economic advantage of solar PV is its modularity at any scale. As Amory Lovins points out in *Small Is Profitable: The Hidden Economic Benefits of Making Electrical Resources the Right Size*, PV can be brought on-line in phases, panel by panel or field by field.[16] This characteristic allows for electricity generation to begin a few months after construction, unlike large-scale power plants (nuclear, hydro, coal, or other), which take years to build and generate no electricity at all until completion. This modularity can dramatically improve the economics of a project, as the electric output can begin sooner and ramp up in proportion to the money spent for that project, generating revenues much more quickly than larger generator installations. Tucson Electric's Springerville facility, for example, has grown by stages—1.35 MW, 2.4 MW, 3.8 MW, 4.6 MW—generating electricity throughout its growth.

Modularity also makes PV systems scalable so that they can be sized to match the smallest to the largest application needs. This scalability is particularly useful for industrial users that need to increase electricity use at a facility in smaller increments than the addition of larger conventional generators would allow. Many firms face difficult decisions when they reach the maximum output of their current generators because deciding to build additional conventional capacity is a large and risky commitment. PV systems will allow for more gradual additions to electric generation, particularly when integrated with electrolyzers and fuel cells. Industrial firms will gain economic advantages because of PV modularity, which in turn will increase momentum for PV adoption in industrial markets.

The Utility Response to Distributed Economics

Utilities are unlikely to sit by and allow their core electricity business to be replaced by distributed PV. In fact, in the United States some utilities have historically resisted net-metering laws and mandatory PV requirements for new construction such as those proposed in California in 2004, arguing that it is unfair to require them to provide grid infrastructure for a customer if much of the customer's electricity is being generated independent of the utility. However, this objection is disingenuous because in many areas customers pay a monthly flat fee on their bill for connecting to the grid regardless of how much electricity they use. If the pricing is set correctly, the utility should be compensated at a fair market price based on cost per kWh delivered for providing this electricity stand-by and storage capacity.

Some utilities are considering a shift away from the flat pricing by customer class that is common today to develop ways to price electricity throughout the day in line with changing costs. Variable pricing would charge consumers for electricity based on a variety of factors, including the amount of power the customer uses, the time of day they use it, and customer density. This type of variable pricing would charge customers a fair market price for their electricity because the cost for a utility to provide electricity is based on each of these variables. The problem is that consumers in low-density locations and businesses (which use the bulk of their power during peak period) would probably see cost increases under such a pricing scheme. Any change toward higher pricing would accelerate the motivation for this considerable minority of utility customers to switch to solar, an outcome utilities are eagerly trying to avoid.

As alluded to in the previous chapter, peak shaving by utilities can also reduce the entire utility cost structure by eliminating the most expensive loads that set high daytime prices. Each utility would have to look at the benefits that supplementing their portfolio of generation assets with PV would provide based on their actual cost for intermediate-load and peak-load electricity and the insolation available in their market area. A study performing this type of analysis for the New England region was performed by Kate Martin of the Massachusetts Institute of Technology. It found that adding a GW of PV would bring down overall utility rates by 2 to 5 percent for all of the customers in the region through peak shavings.[17] Given the number of people that make up that market, this would represent hundreds of millions of dollars in annual savings.

Regardless of how strongly utilities get involved in deploying PV to meet the challenges posed by distributed adoption, eventually some end users will find it uneconomic to use the grid as an energy-storage solution for their grid-tied systems. Either the cost of PV will drop so far below that of grid power that battery banks become cost-effective or nonbattery storage solutions will become available. The final step in the transition away from centralized electricity generation will be to incorporate domestic fuel-cell generators, such as those that are being marketed in Japan to power homes and that so far have been fueled mostly by natural gas. A complete energy solution for a home or small office building will eventually include a PV array, an electrolyzer to create hydrogen from water, and a fuel cell that recombines that hydrogen with oxygen to produce electricity when required. If properly sized in generation and storage capacity and combined with reasonably efficient end use of electricity, this type of solution could provide 100 percent of many users' electricity needs in a clean, renewable, and (thanks to the lack of moving parts) low-maintenance manner.

Changing energy economics should ultimately prompt utilities to alter their traditional business model. The potential changeover of grid customers to locally generated PV users may very well compel utilities to capitalize on this trend by beginning to think of themselves as a broader class of energy-service providers rather than just owners of electricity generators and distribution companies. Not unlike the telephone utilities that initially resisted the trend toward cellular telephones in the 1980s and 1990s, electric-utility companies will ultimately realize that they have to adapt to changing economic conditions or risk being marginalized. On a practical basis, this change could be beneficial for all parties as local electric utilities enjoy many economies of scale, large customer pools, and ample reserves of technicians, installers, and capital. As a result, they would be ideal distributors and installers of PV solutions by using their expertise to make the new technology cheaper and transparent for users to adopt, much as they do today with centrally generated electricity. These utilities could integrate distributed power into their existing delivery system to maximize the value of their infrastructure as the momentum toward distributed PV evolves. Just like the Baby Bells that initially controlled the regional land-line phone utilities and that now have absorbed the cellular phone industry, electric utilities may find it both unavoidable and profitable to do the same in electricity.

Today, distributed grid-tied photovoltaics are rapidly becoming cost-competitive with various types of grid electricity in some of the largest markets in the world. Continued growth in these early adopter PV markets, driven primarily by the economic decisions of electricity end users, will drive the learning and cost reductions that will open additional markets. In other words, the more PV is produced and deployed, the cheaper it gets, and the cheaper it gets, the more attractive it becomes to end users. The existing grid-based electricity infrastructure is limited in its ability to respond to these changes because using PV to meet their expensive intermediate- and peak-load demands provides economic benefits to them, and they have little control over a customer's decision to deploy PV on their home or business.

While economics is a powerful driver of industrial change, it is not the only factor that people or businesses consider when making decisions on spending money or switching to a new technology. Broader perceptions of future risks and rewards are also factors that can strongly influence today's decision. In the next chapter, other noneconomic forces affecting the rate of actual market adoption of PV are considered.

8

Solar Electricity in the Real World

The prior chapter discussed how the evolving cost-effectiveness of photovoltaic electricity technology is poised to transform the economics of the energy industry in the next decade. However, many additional determinative variables such as public awareness, the effects of volatile fuel and energy prices, and the political will to support deployment of any energy or electricity technology are not predictable or even always quantifiable.

Choosing to install PV happens at the individual level of households, businesses, or utilities. When these decision makers are going through the process of evaluating and deciding to install PV electricity, adoption is driven partly by economics and partly by other factors, including an awareness of PV as a potential solution, the time required to become comfortable with the new technology, an assessment of the risks created by switching from a current type of electricity to PV, and an assessment of the risks of not doing so. Access to the credit needed to finance these systems is also important, particularly in the credit-starved areas of the developing world. As in every industrial transformation, businesses will emerge to provide information about PV and to simplify ancillary financing, maintenance, and risk-mitigation services for PV adopters.

As the electricity and broader energy industries transform, both winners and losers will emerge, creating many social and exogenous benefits but also threatening many aspects of the existing global economic system. Some of the most commonly perceived likely losers in the shift to solar energy, existing fossil-fuel providers and utilities, still have a robust opportunity to respond to future changes in the energy economy and to participate in and benefit from the transition—a move that some are already beginning to make.

What Really Drives Adoption

In terms of the market development of PV technology, as the cash cost of PV decreases relative to the cash costs of generating electricity using conventional fuel sources, the competitive balance will conceptually shift toward PV. However, this shift will not necessarily cause everyone to adopt or even want to adopt photovoltaics. Economic growth often lags behind economic opportunity, and the investment of necessary infrastructure will take time, even as PV technology becomes comparatively cheaper. What can be said with certainty is that PV will become more and more firmly entrenched in the market as it becomes more comparatively economic over time. As long as the costs of distributed PV continue to drop as quickly as or more quickly than the costs of grid electricity, the transition toward PV will proceed steadily, and higher sales and lower prices will form a positive feedback loop. Additional, noneconomic forces will also play a role in determining the rate and locations where PV is actually adopted, and these are discussed below.

Indirect Benefits

Many noncash advantages to a clean renewable such as PV factor into a customer's decision making, even though the economic analysis of the previous chapter ignores many of them. These include reduced pollution, enhanced energy security, and the robust infrastructure that distributed generation creates. These types of positive benefits ensure that portions of society are willing to pay a premium for PV electricity—one factor that is driving the current American market growth for Green Tags.

Green Tags programs, which have been set up by many utilities at the urging of federal and state governments, allow electricity users to pay a voluntary premium to guarantee that the electricity they are purchasing is generated by a nonpolluting technology. Consumers opt to pay a slightly higher rate on their utility bill—0.7 cents to as much as 17.6 cents per kWh, depending on the location and type of generator—for electricity that is certified as being generated from a clean electricity producer.[1] Green Tags abstract the environmental and social attributes of each kilowatt hour of electricity from the electricity itself: the consumers who buy Green Tags may be subsidizing renewable generation in their own area or far away. By the end of 2003, some 235,000 electricity users in America had opted to pay such a premium for their electricity, a fourfold increase from 1999.[2]

Businesses, too, can receive image and marketing benefits that do not show up in direct cost comparisons of energy. Many businesses in Europe and America use their clean-energy programs for corporate promotion or as advertising to attract customers and employees—for example, the green DeutschePost building in Bonn and the BP Beyond Petroleum advertising campaign. Many of these early adopters have been willing to pay a premium for renewable energy because managers at these organizations perceive ancillary benefits from pro-active adoption of clean energy programs, and they factor these noncash benefits into their decision making. Not unlike the shift by cosmetic companies to selling products that were not animal tested, large corporations are beginning to find that environmental awareness is beneficial to their image and may similarly end up as a normative and required cost of doing business. People and organizations that share these values comprise a group of early PV technology adopters that are currently installing PV systems on their locations, including many retail and consumer-product organizations in the United States, such as Whole Foods, Wal-Mart, Coca-Cola, and Frito Lay.[3]

Pressures to increase attention on the environmental effects of corporate practices are also coming from the major banks that provide loans to these corporations. In 2005, three of the world's largest lenders—Citigroup, Bank of America, and JP Morgan/Chase—instituted environmental reviews of loans on industrial projects that were designed to determine the effects these projects have in terms of greenhouse-gas emissions and other environmental pollutants.[4] These new loan-review policies reflect a growing awareness by lenders that corporate clients that do not adequately consider the potential effects of future environmental legislation and market trends risk a loss of competitiveness and credit worthiness compared to companies that do.

Hidden Costs

Beyond the cash costs of installation, maintenance, and financing that are included in the economic calculation of cost per kWh delivered, switching to a new and unfamiliar technology creates additional noncash costs for potential PV system purchasers in time and effort to evaluate such a system—that is, *information costs*. Relevant information costs are not easy to measure because many factors may influence a decision maker's willingness or ability to evaluate a technology switch some people are habitual early adopters of any new technology, others are swayed by the political or environmental aspects of a technology, and some simply have easier access to relevant information. As more PV systems are installed, each new user increases aggregate market awareness, thereby reducing information costs for future systems, and this adoption trend gains steam until a technology becomes mainstream and information costs become a small portion of an adopter's total cost—via implementation of standardized solutions and through a general sharing of awareness and technical knowledge.

Hidden information costs are particularly acute in any switch from a centralized-generation, grid-distributed electricity system to one that includes an increasing amount of distributed power from photovoltaics. A prospective purchaser of a PV system has to evaluate the cost of the system, determine the suitability of a home or building for including PV, locate available system components, find suitable financing, research available subsidy programs, and negotiate with the local utility to connect the system to the electricity grid. Alternatively, purchasing grid electricity from the utilities is the simplest solution for most electricity customers in industrialized countries because homes and businesses are usually already wired to the grid and a phone call to the local utility is all that is required to turn the electricity on or off. The utility deals with the issues of financing the cost of electricity generators and delivering its product to the customer. The utility sends a monthly bill to the customer that covers all the equipment and services that it provides plus a profit margin for its value-added services. In effect, the electric utility is paid to handle all of the complex information and decision making related to operating in the energy business, thus reducing this burden for the end user. While this arrangement is embedded in the existing electric utility industry, to date no reliable source for this type of information and decision-making service is provided to residential solar-power customers, and only on a limited basis for commercial customers.

Even when they have made an investment in time and information costs to evaluate adoption, prospective PV customers face various perceived and real risks that stem from their inexperience with the reliability and intermittency of the system, though many of these risks are eliminated by the methods of tying PV systems to the existing grid. Grid-tied PV systems allow their owners to use the grid, in effect, as a battery in which to store excess energy while the sun is shining and from which to retrieve that energy when it is needed at night. This storage solution offers a simple, cost-effective answer to the need for an on-site battery bank, saving an estimated $1 per watt in installed cost, which can represent a 15 to 20 percent savings over stand-alone PV systems that are not grid-tied. For customers who are looking at such an installation, the grid-tied portion of the system reduces concerns about a failure of the system because customers can access backup electricity from the utility without interruption. Users consider PV a low risk extension of the grid electricity system that they have been accustomed to using all of their lives.

Despite the historic resistance by utilities to grid-tied PV systems, these systems simultaneously create value for both utilities and consumers. During the sunny hours when the demand for and the cost of electricity are highest, electric-utility companies are able to absorb excess PV power from customers—especially from residential customers, most of whom are not at home during the day. The utility can then sell electricity to these customers at night when overall grid demand and the cost of generating electricity are lower. Locating the PV systems on the homes and businesses where the power is ultimately needed (on-site distributed generation) also helps reduce line-traffic loads during peak periods. Every kWh of solar power used on-site is a kWh that the grid does not have to transmit, potentially reducing capital costs to maintain and upgrade the grid.

Solar energy will truly be a viable option, however, only when customers can install PV systems as easily as they can purchase utility electricity—when the special risks, costs, and complexities of adopting PV begin to be handled by new firms that act as energy-service providers. Such companies will represent a new generation of the energy-service companies (ESCOs) that have historically provided energy-efficiency upgrades and alternate generating solutions to customers. Companies such as SunEdison and PowerLight in the United States are beginning to offer commercial customers the conveniences offered by traditional electric utilities, including installing turnkey systems as well as providing off-site system monitoring and long-term maintenance contracts.[5] It is only a matter of time before service levels of this type become more common in residential markets, as well.

Built-in PV

Economic analyses of grid-tied photovoltaics usually assume that systems will be retrofit on existing structures—not the most efficient way to install PV systems. *Zero-energy buildings* and "zero-energy homes"—structures that combine maximum energy efficiency with total on-site generation of all energy and electricity needed—are a more effective solution.[6] When constructing a zero-energy building, the builder integrates energy solutions directly into new construction, reducing energy usage through efficient design and appliances, lowering the cost of installation and wiring (compared to retrofitting), and reducing the need for building materials such as roof tiles or shingles replaced by a PV system at the time of construction.[7] These savings reduce the cost of a PV system by as much as one-third over a similar retro-fit PV installation.[8]

Absorbing the cost of PV into the price of a new home or office building, sometimes at little net additional cost for the builder, can dramatically improve the economics of PV installations by reducing or eliminating electric utility bills for the life of the structure. When wrapped into the mortgage of the new home or building, the need for supplemental system financing is eliminated, and the PV installation can be financed using the cheapest form of long-term financing available—residential mortgages that today offer nominal rates of interest at under 6 percent and real interest rates under 3 percent per annum. In addition, if integrating PV into homes and buildings at the time of construction is made a standard feature, the information cost of evaluating the system for each potential purchaser will be reduced, which will allow builders to profitably provide this feature to their customers.

Off-Grid and Developing-World Applications

Some of the most cost-effective PV applications today that are both profitable and immediately deployable are off-grid. The most basic off-grid uses of PV include remote lighting, roadside emergency phones, meteorological stations, and communications repeaters, all of which store daytime solar energy in a battery for nighttime use. Interest in these devices arises from their pure economic superiority for small, remote applications in which alternative types of power are either expensive or practically difficult to install, fuel, and maintain. It is simply not feasible to power many of these applications using the electric grid or dedicated fossil-fuel generators.

In industrialized countries, homes that are far from the grid make an ideal market for off-grid photovoltaics and have comprised the bulk of the global market for solar modules until the late 1990s. The economic rationale is simple: these homes must either run power lines to the grid or generate their power on-site. Beyond a certain break-even distance, it is more expensive to run power lines than to go solar. Other alternatives such as on-site gasoline or diesel generators can be noisy, require skilled repair, and (for back-country homes) may entail expensive long-distance fuel deliveries—giving the economic advantage to solar power in these cases. Most of the existing solar energy and photovoltaic retailers and distributors in America, including companies such as Real Goods in California, developed to cater to this profitable niche market for remote homes.[9] In addition to PV, these retailers often sell a whole range of related technologies for remote living, including composting toilets and ultraefficient refrigerators.

While remote applications are a small percentage of the total energy demand in industrialized countries, in the developing world they are the rule. In Latin America, the Caribbean, East Asia, and the Islamic world, more than 10 percent of the population is without any electricity access, while in South Asia and Sub-Saharan Africa 61 percent and 78 percent, respectively, live without electricity.[10] Overall, more than one-third of people in the developing world, some 1.6 billion people, do not have access to this basic energy resource.[11] These statistics do not tell the whole story because billions more do not have reliable access to electricity and must endure intermittent service, power sags, and power spikes.

All economic growth depends on safe, secure, reliable access to electricity. Unreliable electricity power, particularly in the developing world, disrupts business and is responsible for untold productivity losses. Rolling summertime blackouts in China during 2003, 2004, and 2005 shuttered factories two to four days a week and showed how disruptive this can be to the efficient functioning of industry. For less developed countries, systemic lack of access to electricity can effectively limit a developing society's ability to build its way out of poverty.

Electricity grids represent a highly capital-intensive way to provide electricity to poor and dispersed communities in these developing countries, and diesel generators are often vulnerable to shortages of fuel and skilled maintenance. It is not surprising, therefore, that photovoltaics are becoming one of the most popular methods of providing energy in these locations. The two primary models employed in providing PV in the developing world include *village solarization,* where a large photovoltaic array with battery bank can be used to light common areas and to provide electricity for all types of communications, and *home solarization,* where solar cells on houses provide electricity to individual families.

These PV solutions have tremendous value for those in the developing world who can access them. The benefits are instantly realizable as the use of solar electricity in households for powering lights, radios, television, and recharging batteries frees up time and money that had been spent obtaining candles, kerosene, and disposable batteries. A 2001 World Bank study has found that such expenditures range from $3 to $15 per month for households earning, on average, less than $250 per month—and this does not count time spent collecting firewood or dung for cooking.[12] Another study has shown that among low-income rural households in India, energy can account for up to 50 percent of nonfood spending.[13] Solar PV cells can be a powerful tool to help these families establish economic security.

Many international aid organizations, from the World Bank to specialized organizations such as the Solar Electric Light Fund, have recognized the potential of off-grid PV systems and have been working diligently over the last decade to provide access to solar solutions for the poorest members of the world. However, logistical obstacles exist, of which the biggest is that off-grid PV systems are initially expensive to set up, and loans to help spread out payments over time are generally limited or unavailable. In the industrialized world, mortgages, home equity financing, and third-party loans can spread PV system costs out over twenty to thirty years or more, at reasonable interest-rate levels, allowing industrialized world buyers to avoid large up-front payments and to keep monthly cash payments low. Finding similar credit in the developing world is difficult if not impossible. According to the IEA Task Force on PV for Rural Electrification, this lack of financial services is the largest obstacle to the commercial dissemination of PV technologies in poorer nations.[14]

Lack of basic credit in the developing world is a systemic problem that extends beyond energy issues. Much effort and money have been devoted by nongovernmental organizations to make *micro-credit* available to low-income borrowers. The difficulty of developing these programs is that each area has its own distinct legal structure, language, culture, and political issues. Participants in the process—such as loan officers, collectors, and auditors—must be selected and trained. In many cases and despite overcoming many of these obstacles, underlying legal structure and property-rights issues prohibit any effective progress as war, famine, and political instability destroy decades of work in months or even days. For these reasons, developing adequate micro-credit in the developing world remains a daunting task.

Electricity service deployment in developing countries often suffers from a lack of sales and service infrastructure as well. Even if off-grid PV solutions could be made affordable through financing, these countries need to distribute systems to end users, fix any components that are damaged, and replace batteries as necessary. The lack of basic technical infrastructure has delayed or derailed technology-transfer initiatives of many kinds in the developing world. Solar power is no different, and these obstacles need to be addressed before any program can bring photovoltaics to rural homes and villages—and keep them running.

Despite these obstacles, solar enjoys some clear advantages when compared with other distributed-energy technologies such as small-scale diesel generators. First, sunlight is a universally available resource that coincidentally happens to be most concentrated in many of the least-developed areas of the world. South Asian, African, and South American countries have the least access to the alternative forms of distributed energy or grid-scale electricity, but they tend to get lots of sun. Second, the same solar technology can be used by everyone in these developing countries and across whole regions, which allows for concentrated and standardized training and support—a characteristic that many other forms of distributed energy, such as micro-wind and micro-hydro, do not share. Third, compared to any other modern energy solution, less technical knowledge is required to set up and maintain photovoltaic solutions, particularly compared to the diesel generators that are commonly used. Installation of a small home PV system can be as simple as pointing a module toward the sun and then plugging in the battery or charge controller.

Some of the greatest long-term benefits of solar PV applications are this low maintenance and solid-state nature. As with information technologies that have transformed global society in recent decades, nearly all of the deep technical knowledge required for performance in photovoltaics is embedded in the hardware. Making PV requires people with advanced degrees, but installing, maintaining, and using it do not. Ultimately, the simplicity of operation and reliability of solar PV technology will provide powerful drivers for solar commercialization in the developing world. And its ability to rapidly replicate and disseminate the embedded knowledge of electricity generation will create substantial wealth for people in developing nations.

To accelerate adoption of PV in the developing world, governments and industry can either (1) increase the amount of credit available to help defray the up-front cost of these PV systems, or (2) stimulate increased economies of scale in manufacturing to reduce the cost of PV systems, thereby increasing the ability of users to purchase these systems with current levels of income. As previously mentioned, government and NGO programs have attempted to assist with various forms of micro-credit for systems. Grameen Shakti Bank in Bangladesh, for instance, has attempted to provide micro-finance loans for solar home systems, creating up to three-year loans for systems of an average size of fifty watts.[15] Also some international aid programs in countries such as Argentina, Sri Lanka, Nepal, and China have attempted to deploy small PV systems through increased use of micro-credit, but these programs have had mixed long-term success in stimulating local PV market development because of the unreliability of local regulatory structures and difficulty in expanding these programs profitably without substantial direct aid.[16]

An alternative method to create long-term access to PV in the developing world may be to stimulate its adoption in the industrialized world. Investing in industrial-scale manufacturing and research is widely expected to increase PV's cost-effectiveness and thereby make all PV systems cheaper, including those sold in the developing world. Japan and Germany's efforts in the last decade to stimulate the growth of PV production in their own countries have reduced costs to purchasers in the developing world by half over that period.

Accelerating growth in volumes of PV produced in the industrial economies and the resulting scale economies over time will help to enable lenders, system distributors, and their NGO backers in the developing world to provide more cost-competitive electricity solutions. Stimulating scale production of solar technology through guaranteed volume purchases and production subsidies in industrialized countries' markets may then prove to be the most effective way to electrify the developing world. The faster that growth in production of PV systems brings down system costs in industrialized countries, the more quickly access to modern electricity can be brought to the developing world in a long-term, sustainable, and secure way and provide a locally available alternative to depleting forests and fossil fuels.

Winners and Losers

Thus far, this chapter has looked at many of the exogenous forces that will affect actual PV adoption even as the underlying economics shift toward an increased competitiveness for distributed PV systems. Regardless of the specific timing of when each user or market adopts PV, the global trend toward adopting solar energy will continue until PV system installations become a substantial segment of annual new construction of electricity-generating capacity worldwide. By the second or third decade of this century, most new electricity-generation capacity will likely be in the form of renewable (nonhydro) energy from a variety of sources, and new nuclear and fossil-fuel generators will probably no longer be economic to build owing to the large number of cheaper and cleaner options that will be available to both utilities and end users of electricity.

This outcome could be perceived as undesirable for providers of the existing fossil-fuel and nuclear infrastructure, and they will almost certainly respond aggressively as long as they perceive the development of renewable energy to be a threat to their core business. Conversely, this outcome is apparently beneficial for providers of renewable energy and stakeholders in these companies who can expect to experience double-digit annual growth rates long into the foreseeable future. In some cases, these will turn out to be the same. The fossil-fuel providers and electric utilities have ample time to foresee and to participate in this industrial transformation, as some are already doing. For example, BP is already one of the top photovoltaics producers worldwide.

Yet in industrial transformation, particularly one involving industries of the global scale and importance of energy and electricity, there will always be winners and losers. In the anticipated shift to distributed PV electricity, end users will be the primary beneficiaries as they gain direct access to less expensive electricity from PV than they could purchase from their local utility. The enhanced security—personal, community, and national—that comes from generating and controlling this electricity resource locally at stable prices will have many intangible benefits, similar to those that arose from the decentralization of the information and computing industries—namely, resources that are deployed where they are used, in which excess is minimized, and where the susceptibility to systemic failure is reduced.

Society as a whole will benefit in other powerful ways from solar-energy market growth and adoption. It has been estimated that for each MW of electricity-generation capacity installed using natural gas or coal, one job is created; for wind and biomass, between one and three jobs; and for PV, between seven and eleven jobs.[17] To deconstruct this estimate further, consider that today about 25 to 35 percent of the cost of an installed PV system is presently in the cost of installation labor. This labor component is expected to become a larger and larger piece of the price tag as higher learning rates for hardware bring component prices down more quickly than the cost of installation. While production of PV modules and balance-of-system components will probably be outsourced to low-cost manufacturing locations, installation jobs cannot be exported: they must remain in the locations where the systems will be used. Especially in the industrialized world, where installation involves grid connection, compliance with building codes, and other specialized knowledge, PV installations will become a significant part of building trades, such as electrical, plumbing, and carpentry. These relatively highly paid, skilled jobs will help sustain an educated and prosperous middle class in any industrialized economy.

These benefits will be welcome for consumers in the industrial world, but they will be vital for growth in the developing world, where economic efforts are often paralyzed by inadequate or volatile energy access. Globally, any market development that reduces the price of such a material input as energy will confer economic benefits that ripple through every product or service in proportion to its energy content. As a result, labor and capital productivity will increase, generating wealth for economies while mitigating inflation pressures from rising energy prices.

The economies of the developing world will also benefit from solar-energy adoption because the vast majority of these countries depend on oil imports to fuel local industry. Locally generated electricity will help even out the balance of trade for these countries and provide the opportunity for international aid to remain in-country and provide multiplicative local economic benefits rather than being exported as payments for fossil fuels. Industrialized, but fossil-fuel poor, societies will be able to improve their own balance of trade and energy security, thereby reducing the motivation to intervene militarily to protect access to energy supplies.

Fossil-fuel-rich countries, including OPEC members and Russia, could potentially view a shift away from fossil fuels as threatening. At the same time, many oil-rich countries lie in the highest-sun areas of the world—the Middle East, Africa, Asia, and the Caribbean. Once these countries reach their peak production in oil and natural gas, whenever that turns out to be, they will have to develop cost-effective energy replacements for declining fossil-fuel reserves. It is an interesting coincidence that many of the most oil-rich countries of the world are also the sunniest. The Middle East, in particular, with vast empty desert areas and large amounts of consistent sun, would gain many of the same advantages in bulk PV power that they currently enjoy in fossil-fuel markets.

In addition, new and equally vital industries could benefit from access to globally distributed and inexpensive solar electricity. As discussed in chapter 3, declining water availability is one of the largest problems facing the developing world, proportionately larger in the drier and sunnier parts of the world. Few adequate solutions currently exist to provide additional water supplies as underground aquifers continue to be depleted and freshwater in lakes and rivers is increasingly diverted. The problem of water availability is made more difficult by the economics of water distribution. Water is heavy and, despite its vital nature, of relatively low economic value for its weight, making it economically prohibitive to transport over long distances, which is why most water solutions have involved local ground pumps as opposed to pipelines or trucking. The most promising (and in some cases only) solution that nations have employed to provide freshwater to islands and other remote locations has been desalination. Unfortunately, most of the cost of desalination, either via thermal distillation or reverse osmosis, is in the energy used during the desalination process, limiting its economic deployment around the world. PV can be used to power either thermal or reverse-osmosis desalination plants.[18]

Declining PV prices make these projects increasingly feasible, and the necessary sea water and solar power to run them are ubiquitous and often available in the same location. With more than half of the people in the world living within sixty miles of an ocean, PV can become a powerful tool to facilitate access to adequate water supply.[19] Increasingly cheap generation of solar electricity has the potential to provide affordable freshwater without any need for batteries, power lines, or fuel supplies—and can do so with modular systems ranging from small domestic to large industrial.

Finally, cheaper electricity will drive cost reductions in fueling global transportation, including automobiles, trucks, trains, planes, and ships. In fact, many early automobiles were powered by electricity, and by 1904 nearly one-third of all cars in Boston, New York, and Chicago used this technology.[20] At the time, the relative high cost of electricity versus gasoline combined with the longer range of internal combustion motors initially gave the economic advantage to gasoline-powered vehicles. In the interim, the entire petroleum infrastructure that developed to power these engines—refineries, tankers, and gas stations—has continued to reinforce this economic advantage, and attempts to reintegrate motive and stationary power applications (that is, to power vehicles using stationary power plants) have since been frustrated. The last mass-production attempt to produce the battery-powered electric car, the EV program in California launched by General Motors in 1996, fizzled as a result of the short driving range and long charge times for these vehicles. Environmentally, an electric car is only as clean as the plant that generates the electricity it uses, which is still primarily fossil fuel-based for most industrialized electricity grids. Today, the next generation of electric vehicles running on internal hydrogen fuel cells offers a new potential development path, combining the benefits of using clean renewable electricity with the range and power of fuel-based engines.

Despite the optimism of advocates, industry groups, and governments, the transition to hydrogen for use in motive applications will take decades due to serious issues involving hydrogen supply. Contrary to popular belief, automotive companies already know how to make electric drive trains cheaply and effectively. Although the California EV program was ultimately unsuccessful in mass deployment of electric-powered cars, one of the enduring benefits of the attempt was the development of electric drive-train technology to a level of cost and performance completely adequate for today's uses. Today, the real obstacle to wider deployment of hydrogen fuel-cell cars remains how to provide the hydrogen to run the fuel cell consistently, economically, and cleanly.

Leaving aside several economically and environmentally questionable schemes that would add on-board reformers to allow fuel cells to run on fossil fuels, three possible solutions to the hydrogen problem exist: (1) finding a way to deliver centrally generated hydrogen from nuclear plants directly to end users (similar to today's delivery of propane to homes and businesses), (2) creating hydrogen fueling stations similar to (or integrated with) today's gas stations, and (3) encouraging local generation that uses electrolysis or direct solar energy. With the first two solutions, centrally generating hydrogen and delivering it to homes and fueling stations would require a large generation and transportation infrastructure that would be costly to create and that must be in place before fuel-cell cars can be widely deployed. However, automotive companies are reluctant to invest capital in designing and manufacturing fuel-cell vehicles until they are more certain that the necessary hydrogen fueling infrastructure is in place. This dilemma cannot be escaped without massive capital expenditures and possible government involvement.

Under the third possible path, fuel-cell cars will be fueled through the production and storage of hydrogen where cars are used and parked—a capability that will be widely available as a side-effect of the growth of photovoltaics, electrolyzers, and fuel cells for stationary electricity generation. As prices for PV and small fuel-cell generators drop, homes and offices using these systems can also have electrolyzers on-site to create hydrogen. Distributed PV systems are the only renewable technology that can create electricity and therefore hydrogen on such a micro-distributed basis. When stationary fuel cells become commonplace, hydrogen will be available at the locations that cars are most often used to access—homes and businesses. Long-distance car travel will still require building some hydrogen fueling stations, but the existence of a large distributed network of hydrogen generators already in place in homes and businesses would make the transformation simpler.

Regardless of the path that is taken to breaking the dependence of the modern transportation infrastructure on oil, mobilizing fuel cells in cars and trucks ultimately will rely on the deployment of renewable energy and the clean, local, and cheap generation of hydrogen.

For many of the reasons described in this chapter, the adoption of PV can be either enhanced or inhibited by the information available to customers when evaluating a potential switch to this new technology. Beyond the opportunities for businesses to step in and fill the information gap by developing new distribution business models, it is the role and responsibility of government to get involved where social welfare can be efficiently improved. Many governments around the world, from Europe to Asia to North America, have already determined that accelerating the adoption of PV electricity will provide their citizens with significant environmental and social benefits. These governments have created and deployed various incentives to help overcome users' remaining perceived barriers to wider PV deployment and to improve system economics further. Many of these tools are explored in the next chapter and collectively provide a portfolio of options for governments at all levels to more quickly capture the inherent economic and environmental benefits that PV electricity provides.

IV

A Promising Destination

9

Tools for Acceleration

Economics is driving the energy and electricity industries to develop more renewable-energy technologies, which will also create ancillary wealth, security, and environmental benefits around the world. These social benefits are potentially vast and should be encouraged through progressive government policies as well as coordinated industry efforts. Thoughtful investments in attention and money made today by industry and government are highly likely to accelerate this change and bring dramatic financial and social returns.

Various policy tools have been created and used at all levels of government from local to international to help accelerate the adoption of renewable-energy technologies, including PV. Without the implementation of such policy tools in Japan and Germany over the last decade, the PV industry would not be enjoying its current rapid growth and market opportunities. Today, social and political pressures coupled with rapidly rising fossil-fuel prices are increasing the motivation for most jurisdictions around the world to evaluate additional solar-energy industry support programs, including rebates, feed-in tariffs, and R&D support programs. In addition, the private sector is developing coordinated and collaborative efforts by industry players to standardize equipment and connection methods, educate PV system installers, and access capital markets, steps that all young industries take in their progression toward broad adoption and market growth.

A Level Playing Field?

The conventional energy industries, including fossil-fuel and nuclear-energy providers, are among the most heavily subsidized in the world, second only to transportation, an industry intimately linked to energy as well.[1] As a result of this governmental subsidy and legislative support, the market prices of conventional energy and the electricity generated from them do not currently reflect the real cash costs of their production, much less the external costs inflicted on wilderness, farming, forestry, and health.

Global government support is currently skewed toward the nuclear and fossil-fuel infrastructure, with about ten times as much money going to these conventional power sources as to all renewables combined. Subsidies also exist for electricity grids, particularly in the developing world, where retail electricity prices are kept artificially low through cheap government-guaranteed financing and direct payments. In addition, the transportation infrastructure, including roads and highways, can also be thought of as an associated subsidy to the fossil-fuel industry and in particular the oil industry, which was a member of the highway lobby that helped pass the Federal Aid to Highways Act of 1956 in the United States, the largest public-works program in U.S. history.[2] The economic effect of the existing regime of energy subsidies, not unlike all subsidy programs, encourages consumers to use fuel, electricity, and transportation more inefficiently than if they were paying for its development and upkeep directly.

Worldwide, the fossil-fuel and nuclear-energy industries receive direct subsidies totaling $131 billion every year.[3] These estimates count only cash subsidies, ignore any indirect benefits (including reduced-cost financing for energy producers, tax subsidies, and extra military and police spending to secure supplies of fuel), and ignore reduced environmental-protection constraints that would increase the estimate of subsidies to these industries by another $200 billion per year.[4] Indirect subsidies can total many hundreds of billions of dollars more. In contrast, governments worldwide spend less than $5 billion combined on renewable-energy subsidies every year, which equals only 3 to 4 percent of the direct subsidies and 1 to 2 percent of the total subsidies given to conventional energy providers, despite the environmental advantages that renewable technologies provide.[5]

Promoting PV

Because of the conventional energy industry's entrenched political influence, public subsidies have historically been difficult to eliminate. One analysis estimates that between 1993 and 1996, American oil and gas companies made political contributions of $10.3 million and received tax breaks worth $4 billion.[6] Reversing many longstanding fossil fuel subsidies could greatly accelerate the transition to renewable-energy technologies. Until that occurs, unsubsidized renewable-energy technologies must be even more economic than subsidized fossil fuels to achieve competitiveness and market adoption.

If reducing or repealing existing fossil-fuel subsidies is politically difficult, other methods of leveling the playing field for renewables can be employed. Two basic strategies have emerged: (1) create incentives for the production and installation of renewable-energy alternatives, and (2) penalize extractive and polluting fossil-fuel energy sources. Other methods of stimulation, including government-sponsored research and development programs and streamlined regulations for connecting grid-tied PV, can also help to increase their adoption.

Incentives for Adoption

Programs and incentives that have directly stimulated the recent growth of renewable-energy markets and photovoltaics, in particular, include feed-in tariffs, net metering, rebate programs, consumer tax deductions, and production tax credits. All of these provide the technology buyer either a direct reduction in the up-front system cost or a payment for the value of the energy that the system produces over time. Different countries have used different combinations of these incentives to promote the growth of their domestic renewable-energy markets.

Feed-in tariffs requires a utility to accept power from a renewable-energy source at a set rate over a certain time period, sometimes as long as twenty years. Rates may differ based on the type of renewable energy, size of the installation, and time of day and year—and are usually measured in cents per kWh. The laws establishing feed-in tariffs typically set forth the right of the owner of the renewable generator to connect to the grid and also establish a standard method for doing so. Most feed-in tariffs are structured to provide payments to installers, ensuring long-term revenue streams that compensate them for the amount of energy they produce.

Some countries and U.S. states alternatively use cash *rebates* to lower users' out-of-pocket costs for adopting renewable, primarily PV systems. Rebate programs usually pay consumers some cost per peak kW on certification of the finished system. Since buyers receive cash payment at the time of installation—up to half or more of system cost—their economic risk is reduced since it no longer at risk of potential shifts in government priorities. Despite a lack of federal programs in the United States to support the PV installation, many states have elected to institute rebate programs to support PV adoption. These programs dramatically reduce the initial cost of PV system installation and collectively are the primary force propelling the growth of the U.S. photovoltaic grid-tied market.

Another instrument that governments can use to change the market dynamics of renewable energy is tax policy. *Investment tax credits, producer tax credits,* and *consumer tax incentives* comprise a class of incentives that can be used to stimulate volume manufacturing and are variously used by many states and, since mid-2005, by the U.S. federal government, which instituted a two-year 30 percent tax credit for PV systems. The goal is to reduce the out-of-pocket costs borne by the owner of a PV system by reducing the taxes that owners pay for clean-energy systems or efficiency improvements. Tax credits can be issued that can be either captured directly by the installer or sold for cash to outside investors looking to shelter income.

Penalizing Pollution

Another way to increase the competitiveness of renewable energy is to cap carbon emissions produced by fossil-fuel energy sources through *carbon trading rights*. Modeled on the U.S. Environmental Protection Agency's successful Allowance Trading programs for reducing sulfur dioxide and nitrogen oxide emissions of coal-fired power plants in the 1990s, which helped to decrease acid rain and particulate pollution in Canada and the United States, carbon-trading programs (such as those embedded in the Kyoto Protocol) hope to replicate this success with fossil-fuel carbon emissions.[7]

The objective of carbon-trading rights is to determine target emissions of carbon dioxide (and other greenhouse gases) by country and then to allocate or sell fixed emission rights to existing producers within each of them. These producers could then either implement technical fixes to stay within these emission restrictions or could purchase the rights to produce emissions from other holders in an open market, with sellers reducing their annual emissions accordingly. Such a cap-and-trade (cap total emissions and trade rights to emit) system allows the energy industry as a whole to allocate pollution-reduction initiatives to the locations and applications where those reductions can be most cost-effectively achieved. To meet the targets laid out in the Kyoto Protocol, which came into force in 2005, the twenty-five states of the European Union are attempting to reduce carbon emissions to 8 percent below 1990 levels by 2012 by launching carbon cap-and-trade schemes in 2005, with prices for these emission rights quadrupling in the first six months of trading.[8]

Alternatively, governments can use tax policy to penalize pollution or the use of technologies that contribute to global climate change. *Carbon taxes* or *emissions taxes,* not unlike the existing gasoline and diesel fuel taxes of many industrialized countries, can be used to penalize emitters of pollution and greenhouse gases and affect how much and what types of energy they consume. Carbon taxes, by charging a tax for each unit of carbon emission, create incentives for polluters to clean up their production methods. Tax methods do not cap total emissions as cap-and-trade schemes do, limiting the ability of governments to manage aggregate environmental impacts, but they can be easier to administer and represent an additional method that governments can use to change the relative economics between polluting fossil fuels and renewable-energy technologies.

Research and Development

Aside from directly affecting system costs and revenues through subsidies and tax breaks or adding additional cost to competitive polluting fossil fuels, the method most commonly used by governments to support new energy technologies is research and development funding (R&D). In the United States, the Department of Energy spent $212 million in 2004 for renewable-energy R&D primarily through the National Renewable Energy Laboratory in Golden, Colorado.[9] R&D funding by industrialized countries' governments for renewable energy is crucial for market growth because it helps resolve a commonly observed market failure in economics—that is, businesses collectively underinvest in R&D and basic science compared to what a socially optimal level would be. To compensate, governments of industrialized nations often support basic research on promising technologies. Energy R&D investments, in particular, yield benefits far in excess of their cost. A study by the U.S. Congressional Budget Office shows that over a twenty-two-year period $7 billion in U.S. Department of Energy investments in more efficient use of energy generated some $30 billion in benefits.[10] However, despite the clear payoff from R&D investments in efficiency improvements and renewable energy, annual global energy R&D spending dropped by almost two-thirds between 1979 and 1996, disproportionately in renewable-energy R&D, due to low fossil-fuel prices and changing geopolitical priorities.[11]

Enabling Market Access

To capture the economic advantages of grid-connected distributed PV electricity, system purchasers must be able to establish *interconnection* of their systems to the local utility grid. Often, utilities have specific and complicated connection requirements, equipment checking, and fees for processing the interconnection application. This situation is particularly acute for installers in the United States, who may have to install systems across various states or utility coverage areas, each of which may have different requirements.

Once customers can connect their systems to the grid, they need to be properly compensated for the energy that they produce and send into the grid. *Net metering* laws can be set at any rate for the energy that passes between the customer and the utility, but most often they pay a fixed rate for energy flowing both ways, thereby functioning as a true net energy exchange. As mentioned, utilities that charge customers on a fixed-rate structure for electricity gain an economic benefit under net metering, taking in expensive daytime energy from the customer and reducing it at cheap nighttime costs.

Perhaps the most effective policy that governments can put in place for developing the solar industry is to mandate that increasing numbers of new homes and commercial buildings include solar components in their initial construction.[12] By requiring architects and builders to consider the energy characteristics of building structures and to install PV during construction, the cost of PV systems can be dramatically reduced compared to primarily retrofit market. In addition, subsuming the financing of a PV system into the mortgage of the home or building would make paying for a PV system simple and provide the most favorable interest rates.

New-construction mandates can also provide PV manufacturers with predictable sales, a market development that would motivate them to build larger-capacity plants that in turn would decrease PV costs even further. In 2004, the California legislature debated such a proposal to mandate 50 percent of new-home construction in California be solar-integrated by 2010. The initiative was ultimately defeated in the California legislature but signaled the increasing awareness by governments that PV electricity installed at the time of construction can provide cost-effective renewable energy.

PV Policies around the World

Many jurisdictions around the world, from cities to states to nations, are evaluating or implementing the policies discussed so far in this chapter to stimulate adoption of renewable energy and electricity and the growth of local producers and installers. Japan and Germany are among the most proactive, though a number of other locations, including many U.S. states, Spain, and China, are also aggressively pursuing such strategies.

Japan: A Policy Success Story

Governments often stimulate the growth of alternative energy by using a portfolio of approaches to stimulate the different components of a technology supply chain. In the 1990s, the government of Japan developed such a portfolio of tools and programs to stimulate different components of the PV technology supply chain—from production to installation—that have allowed photovoltaics in Japan to attain cost-competitiveness with grid electricity over the last decade. This model of success has created the foundation on which the worldwide industrial transformation toward distributed PV electricity is based.

With scarce domestic energy resources, Japan has experienced high electricity prices for much of the last century. Today, half of Japan's primary energy comes from oil, 85.5 percent of which is imported from the Middle East.[13] The oil shocks of the 1970s and a cultural and geography-ingrained preference for independence prompted the Japanese to develop domestic energy sources, including nuclear energy, which now generates over a third of Japanese electricity. The Japanese government has also spent relatively more than other industrialized nations on R&D for renewable energy, particularly photovoltaics. With broad support from universities, local governments, and businesses (including its highly integrated *kieretsu* holding companies), the Japanese government has sought to create a domestic manufacturing base to provide low-cost PV solutions for both domestic electricity generation and export markets. Programs have included PV-targeted subsidies for schools and public buildings, support for regional government programs (which orginally paid up to half the cost of domestic PV systems), and incentives for businesses to adopt PV.

The start of the Seventy Thousand Roofs program in 1995, however, accomplished the most for Japan in terms of dramatic PV industry growth and a halving in the cost of PV systems since that time. The Seventy Thousand Roofs program initially provided for a 50 percent subsidy on the cost of installed grid-tied PV systems. By setting subsidy levels so that the net electricity cost to the customer was competitive with conventional electricity options, the program prompted rapid growth of the PV market and spurred supply-chain development by manufacturers, integrators, and installers. As a result, the unsubsidized price of PV systems in Japan has fallen from $11,500 per peak kW in 1996 to a little more than $6,000 per peak kW today. With appropriate financing, these systems are now providing electricity at or below the average residential rate of $0.21 per kWh.[14] As increasing industry maturation has brought down prices, subsidies have been tapered off to keep the buyer's out-of-pocket cost essentially equivalent with conventional electricity options. Figure 9.1 shows the change in both the unsubsidized and the subsidized prices of PV in Japan since 1994 and the rapidly increasing number of applications for funding under the government subsidy program.

Japanese success in driving costs out of PV systems is not surprising, given that since World War II, the Japanese have developed extensive expertise in cost engineering and in precision mass manufacturing in the consumer electronics, computer, and automobile industries. Microelectronics producers such as Sanyo, Sharp, and Kyocera are leading the charge to drive cost out of PV manufacturing as a natural extension of this skill base. Japan has also enjoyed some of the cheapest capital in the developed world over the last decade, with nominal mortgage rates just above 2 percent and real rates approaching or below zero.

The Japanese residential PV program expired in 2005, having achieved its goals of making PV cost-competitive with conventional electricity

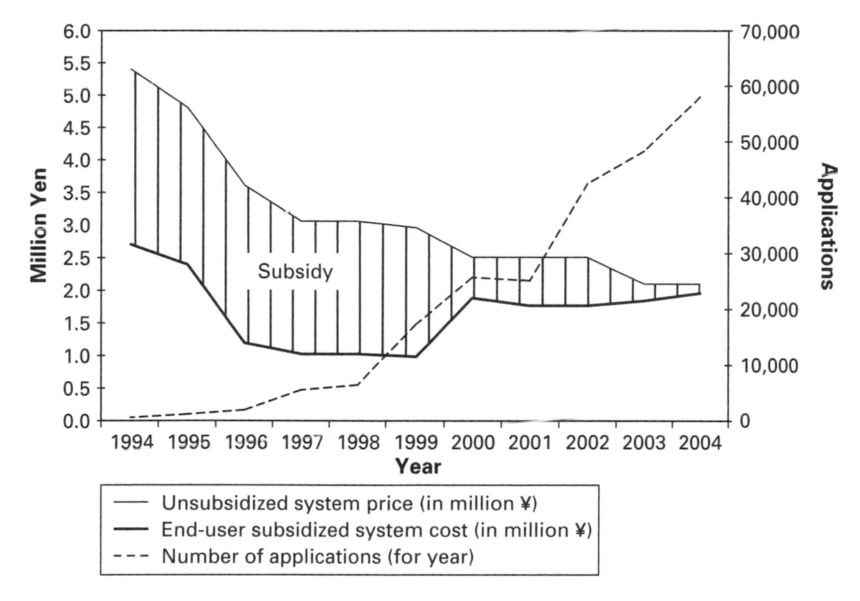

Figure 9.1
Subsidized and unsubsidized PV system costs in Japan and annual applications for rebates, 1994 to 2004.
Source: Photon International (2004).

options and building a solid base of manufacturers and installers.[15] It remains to be seen whether the program created a market that can sustain itself and that can maintain its recent historic growth rate of 30 percent per year.[16] Japanese PV manufacturers seem to believe it will, and Japanese producers increased domestic production of PV cells by over 65 percent from 2003 to 2004.[17] Japanese PV manufacturers continue to expand production, increase exports, and set up operations in places from China to Brazil to Mexico. Clearly, the Japanese government's progressive PV policies of the last ten years have firmly established Japan as the world leader in PV technology.

Germany and Europe

In Europe, feed-in tariffs have been used successfully to stimulate renewable-energy-market growth for both wind and PV. Originally instituted in Denmark, currently the largest and most successful feed-in tariffs program for solar energy and wind power is found in Germany. Beginning in 1990 with its Thousand Rooftops program and supported by a number of local-level initiatives in the mid-1990s, Germany began to develop its PV industry infrastructure. From 1999 to 2003, the Hundred Thousand Rooftops program, championed by the German Green Party, generated huge interest in PV systems by offering a 50 eurocent per kWh feed-in tariff on installed systems for twenty years plus low-interest financing. As a result, installed PV capacity in Germany tripled from 41.9 MW in 1997 to 113.8 MW in 2000[18] and again to 385 MW in 2003.[19] In early 2004, the German government renewed the feed-in tariff program, and the market for PV installations in 2004 grew by over 150 percent from 2003.[20]

Not only has Germany been interested in promoting PV, but it has begun phasing out subsidies for coal and nuclear power. Germany is phasing out subsidies for construction of new nuclear power plants and is also planning to decommission its existing nuclear power plants. According to a negotiated settlement between the government and the utilities in 2000, the last German nuclear power plant will close in around 2020.[21] The German government is counting on robust renewable-energy supplies, including PV, to help fill the electricity gap that decommissioning its nuclear plants will create. Despite tight supplies in the silicon supply, manufacturers are also optimistic and are expanding their module production in 2005 by nearly 50 percent over 2004.[22]

Many other governments around the world are also establishing programs based on the success of the German law. For example, China's renewable-energy framework was established in February 2005 and modeled after Germany's, and the Chinese law is already spurring an increase in domestic PV manufacturing.[23] In a country where manufacturing capacity is idled for days each week due to lack of supply during peak summer electricity demand, PV could and will fill an important role in supplying such electricity. The success of the German feed-in tariffs program has also spurred other European countries to implement or to consider implementing these programs. For example, Spain has had feed-in tariffs for PV since 1998 and is now the second-largest adopter of photovoltaics in Europe.[24]

The United States

Beyond the research and development of the National Renewable Energy Laboratory, direct subsidies for PV in the United States have historically been left to the responsibility of state governments to prioritize and fund.

While the federal government has had a production tax credit for windpower for many years, it was only in 2005 Congress passed a two-year 30 percent tax credit for residential PV systems as well.[25] Though this was a small piece of a very large energy bill, the inclusion of a PV tax credit shows that the U.S. government is again taking solar energy seriously as an option for providing local renewable electricity.

Still, the states are doing most of the work in promoting PV. The state and municipal governments of the United States do so most often using rebates for system purchases to spur adoption by customers, with the exception of a new European-style feed-in tariff established in Washington state in 2005. Figure 9.2 shows the kinds of rebate programs that are offered in the United States at the state level. This figure ignores limited regional or local programs such as the Tucson Electric program mentioned in the previous chapter, which offers up to $3,000 per kW; Sacramento's municipal Green Power program, which offers roughly $3,500 per kW; and Los Angeles's program, which targets 100,000 solar rooftops by 2010 and offers $3,500 per kW (with $4,500 for systems that are manufactured locally).[26]

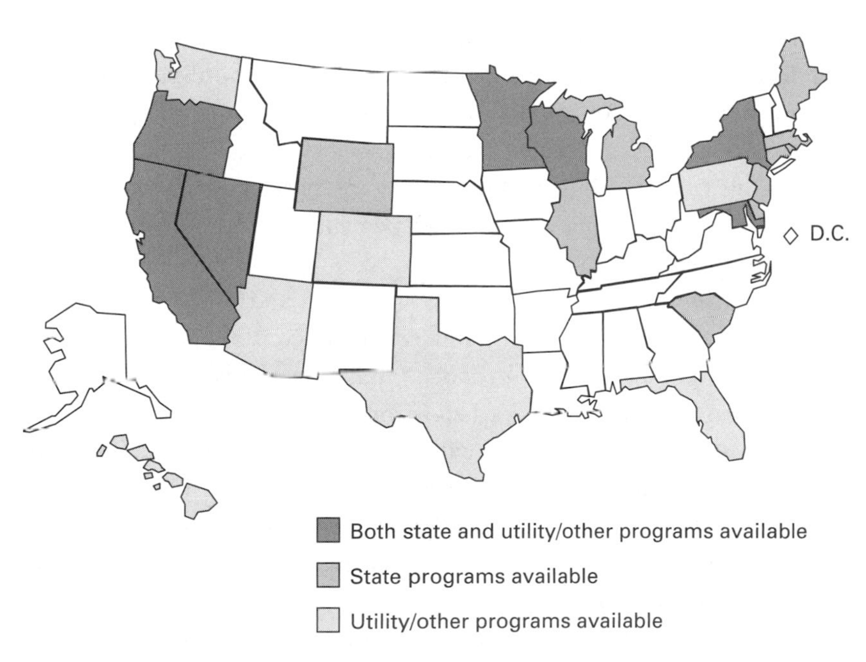

Figure 9.2
U.S. states with rebate programs for renewable energy technologies, October 2005.
Source: DSIRE.

Renewable portfolio standards (RPS) programs have recently become more prevalent in many states' efforts to deploy clean energy and electricity-generation technologies. RPS programs are attempts by states to set target percentages of renewables in their power-generation mix to be deployed by certain dates. For example, Colorado passed a RPS in late 2004 that mandated that its electricity producers generate 10 percent of their electricity from renewable sources by 2015. Collectively, twenty-eight states have instituted RPS laws already, and this number will likely grow.[27] A potential pitfall for PV in these programs is that the renewable contribution requirements are often met by centrally generated electricity, predominantly windpower. A few RPS programs, such as those in Colorado and Pennsylvania, make explicit carveouts for solar-energy contributions to the targets, ensuring that the programs will promote a wider portfolio of renewable energy options from which states may draw.

Private-Sector Initiatives

Aside from government programs to stimulate the increased use of PV, the solar-energy industry itself is taking steps to accelerate adoption. In establishing any new technology, it takes time to coordinate industry growth because various mechanisms, institutions (including product standards, training, and certification), and specialized capital markets must evolve to reduce information costs to potential customers. To keep the PV industry growing at 25 to 30 percent per year for decades to come, these must continue to be developed.

Industrywide product-performance standards and rating scales must be agreed on and held to a high standard so that consumers who are contemplating a purchase of PV products and services know what performance they can expect. Standardizing connection of PV systems to the grid at the state or national level will allow potential users to know that they can easily grid-tie their systems at a minimal cost or risk.

National certification for PV installers is also critical because the workers who perform installations must be technically competent and must be regarded as providing reliable service. In other words, a buyer needs to be able to trust both what comes in the box and who wires the connection. Many PV distributors in the United States provide training programs to customers who wish to install PV systems for themselves or others. The North American Board of Certified Energy Practitioners has instituted a standards and training program for PV installers and hopes to set the standard for qualified installers in the years ahead, but more must be done to create national or international standards for professional installers and the people who train them.[28] The same need for training applies to architects and builders so that they can properly specify and construct energy-efficient buildings with integrated PV. As mentioned, the integration of energy-smart features into the design and building process reduces both PV product and installation costs, and this training will help architects and builders explain the economics and value of PV systems to their clients.

Finally, having adequate access to financing at reasonable rates over appropriate time periods is indispensable in adopting a capital-intensive energy solution such as PV. From the customer's point of view, the cost of a PV system includes all the energy that it will supply over its lifetime of thirty years or more. Being able to use the best financing available spreads the high initial cost of the system over the many years that the system generates electricity and helps to match the payments to the benefits received, not unlike a traditional mortgage or a car loan. Lenders who understand the unique risks and characteristics of lending against PV systems and related contracts can help stimulate PV adoption by providing specialized financing to building owners who wish to retrofit existing facilities. Longer loan terms, inflation-adjusted payments, and appropriate values for systems at the end of the loan period will dramatically lower the monthly cash payments for PV electricity as easily as would improvements in the underlying technology.

Aside from debt, access to equity financing for growing PV technology companies is critical. The years 2004 and 2005 saw a surge of venture-capital investments and public-equity initial public offerings for renewable energy, in line with rising fossil-fuel prices. In 2004, $520 million was invested by U.S. venture-capital firms into clean technologies, more than double the share of all venture investments from the year 2000.[29] Eight companies alone are going public in various global equity markets in the second half of 2005, a huge number considering the small size of the global PV industry, indicating serious long-term interest by investors.[30] Research analysts, equity funds, and investment bankers will capitalize on evolving business opportunities based on their perception of the current and potential future growth in value of the PV sector, but the momentum is clearly toward more mature involvement by global capital markets in PV companies and technology.

This chapter has reviewed many of the methods by which governments and advocates of the PV industry use public and private institutions to stimulate the industry's growth. Going forward, small amounts of government support and subsidies have the potential to create large long-term social benefits by accelerating PV adoption in the next decade—in the process providing a quicker realization of the scale economies in PV production that will pay economic and social dividends for decades. Industry, too, should further accelerate the establishment of standards for components and training for designers and installers to ensure that consumers have ready access to the knowledge needed to make informed decisions and implement PV solutions. These will expedite the transition to and growth of the PV market, and history shows that the faster that such institutions can be developed, the quicker the deployment of the technology will occur.

The various drivers of the coming transition to a dominant role for PV in our global energy industry have been developed and discussed in the last few chapters, including the economic, noneconomic, and institutional forces driving its evolution. The next chapter takes a broad view of how all of these forces will interrelate over the next few decades.

10

Facing the Inevitable

This book has looked at the history of energy, present circumstances, and the drivers of the revolutionary changes that are going to transform the global energy and electricity industries in the years and decades ahead. This final chapter synthesizes these various forecasts and trends into a comprehensive view of the transformation that they imply, the phases that will be experienced in succession as the transformation unfolds, and the potential limits to growth that all transformations must ultimately face.

The transition described in this book is the culmination of two key economic drivers—(1) a necessary and desirable shift away from fossil fuels as they become increasingly scarce and damaging to the environment and (2) a shift in the type and location of sources of energy and the generators they use to transform energy into more useful forms such as electricity away from centralized sources to more distributed ones. Figure 10.1 shows how the combination of these two drivers is leading to a distributed renewable architecture in which solar energy and PV are the predominant and most widely available alternative. As economic factors propel both of these key drivers to converge on a distributed solar architecture, the external benefits of stable energy costs, secure access to vital supplies, and reduced environmental impact of our energy consumption will multiply.

Direction, Momentum, and Headroom

The previous chapters have identified a few key economic and noneconomic drivers for both the traditional energy infrastructure and the distributed energy infrastructure that is poised to supplement (and eventually

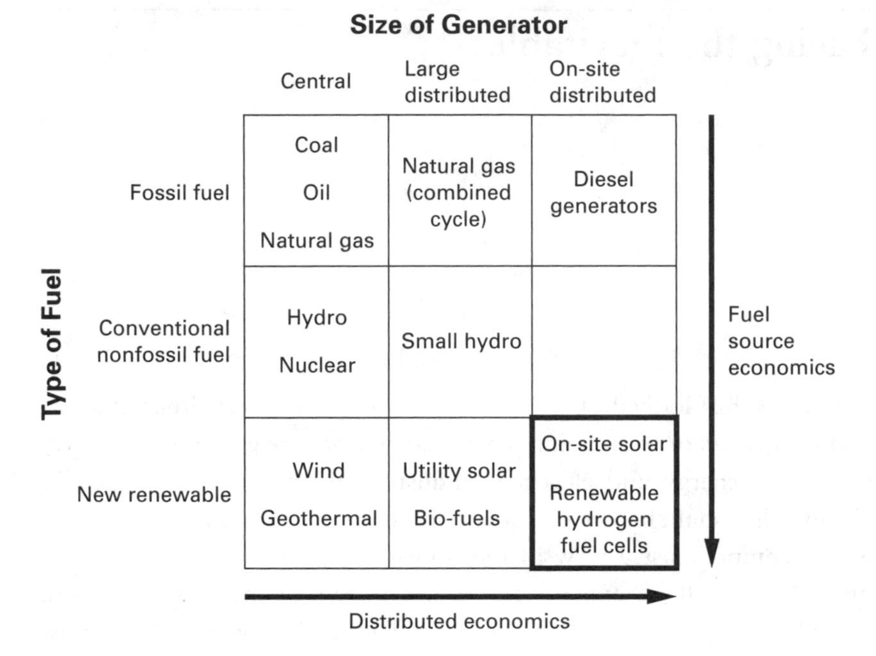

Figure 10.1
The convergence of economic forces promotes a new set of electricity solutions.

supplant) it over the next three to four decades. The most important of these many drivers include a flat to increasing price of grid-based electricity, decreasing cost of PV electricity, and growing awareness of the economic viability and social advantages of deploying distributed photovoltaics. As with any type of forecasting, however, the projected results need to be examined to determine whether it is reasonable to conclude that these trends are both real and long-lasting.

One conceptual framework for understanding long-term trends focuses on direction, momentum, and headroom. *Direction* is the way in which the world moving. In 1900, the direction was toward increased use of fossil fuels, especially oil. Today, it is away from fossil fuels as available reserves are steadily depleted and as the costs of recovering and convering them increases. Wind and solar energies have been increasingly supplementing or substituting for older forms of energy at the margin, and this trend will accelerate for many of the economic reasons described in prior chapters. Socially, many governments and businesses are realizing the need for reliable access to local, nonpolluting sources of energy and electricity and are instituting policies to encourage more rapid adoption of them. All of these forces point to a world of increased demand for and therefore supply of local renewable energy. In 2003, some 10 percent of new electric-generation capacity installed worldwide was nonhydro renewable. Using the economic forecasts of this book, by midcentury, nearly all new energy capacity will overwhelmingly consist of new renewables of solar, wind, geothermal, and biomass energies.

Momentum refers to how solidly a trend is moving in a given direction. *Momentum* is not the same as how quickly a trend is moving. It more accurately describes whether the trend is driven on a fundamental level by basic forces that are going to be maintained over a long period of time. For example, the trend in the United States toward building nuclear power plants once showed fast movement in a definite direction, with rapid growth in cumulative plant orders from to 1965 to 1973. However, this trend had no momentum as plant orders fell off, and by 1978 the trend had dissipated in the United States. Other trends, such as the trend toward stabilizing world population, can be slow yet steady, exhibiting transformative momentum despite their slow pace. Still others, such as the trend toward wind power, may be subject to bouts of boom and bust until they have stabilized but are assured of momentum, at least for a time, by basic technological and economic forces. The trend toward PV adoption may also experience some ups and downs until it achieves a widely recognized comparative advantage in unsubsidized cost in sufficiently large markets. However, the momentum toward declining PV costs as a result of increasing scale and learning is strong and will be the dominant factor in the future of this industry. Its determining factors are not ideology, fear, or even wise foresight; they are profits. Solar power will be increasingly big business because it will be increasingly good business.

Although continuing government involvement should not be relied on, their potential impact should not be dismissed because the world is experiencing another trend with strong momentum toward public promotion of renewable-energy solutions and protecting the environment from carbon emissions and resource insecurity. Using government power to create markets for renewable energy is a particularly strong trend in Europe, where nearly every country is considering or already implementing such policies. Japanese government subsidies, designed to act not as a crutch but a cradle, have permanently changed the global PV picture in just a few years. As previously discussed, many U.S. states and municipalities are also participating in this trend. As the next few decades unfold, therefore, public support for renewables is likely to continue to increase.

In addition, the various sources of solar momentum are correlated—that is, the growth of one amplifies the others. For example, the increasing cost-effectiveness of PV encourages governments that are looking for clean and secure energy solutions to stimulate the solar market further through subsidies, a feedback mechanism that stimulates additional growth. As European countries and U.S. states and municipalities see PV become cheaper every year, subsidy programs become more feasible and effective. New subsidies in any location feed the global marketplace for PV market, stimulating growth in factory size and sophistication and bringing down costs further. As the market grows, PV is increasingly perceived by buyers and investors as a viable real-world option rather than as a mystifying and exotic spinoff from the space program.

Finally, it is helpful to think about any technology trend in terms of the amount of *headroom* it has for growth—that is, the natural limits or obstacles that will eventually slow or halt growth. All trends, even long-lived powerful trends, have limits, and the transition to a solar economy is no exception. Some eventual threshold limit, for example, must exist in driving costs out of the technology. Even the long-term forecasts developed in chapter 5 show a lower limit in the two- to four-cents-per-kWh range on solar retrofits or those installed as part of new home or building construction. Experts such as Martin Green of the University of New South Wales suggest that it is unlikely that total installed costs can ever be trimmed to this level using the current silicon-based technology, but many other experts disagree. Green's estimates project that silicon PV modules, in large-scale production, might be reduced as low as $2 per watt, which would reduce the price of its electricity to eight to twelve cents per kWh on retrofits.[1] To obtain lower costs, alternative PV technologies such as tandem (stacked) cells, printable cells, plastic cells, or chemical cells would have to be deployed.

However, forecasts of the growth in PV developed this book assume that silicon will continue to be a dominant player for the next couple of decades. Technological breakthroughs are possible and even likely within the twenty-year timeframe until these projections suggest that such cost levels will be reached. In January 2005, for example, Canadian researchers announced the development of sprayable, thin-film solar cells that exploit the infrared part of the spectrum that is invisible to the eye but that can be felt by the skin as radiant heat. Since all previous solar-cell designs have been restricted to the visible part of the spectrum, this discovery could perhaps make it possible to produce cheap plastic PV cells with conversion efficiencies significantly higher than today's 6 percent ceiling (eventually as high as 30 percent) in only three to five years.[2] Only time will tell whether this or any other breakthrough announcement is going to bear fruit on the production line, but work is progressing on so many fronts that technological leaps are likely to occur.

Eventually, the volume of PV that is forecast to be deployed will make it an increasingly, visible energy source. Human beings have historically been averse to seeing their energy being made and have often sequestered their home heating systems in closets and basements and their electricity-generation plants away from population centers. As mentioned in chapter 4, modern industrial wind turbines are experiencing some of this local resistance to their location and operation. Similarly, a large scale of PV deployment would make solar panels and arrays a ubiquitous, and perhaps undesirable, visual presence in modern societies. However, PV technology has the additional potential to be integrated directly into building materials, such as roofing materials, architectural glass, and potentially paint and plastic casing. Already, many manufacturers are cost-effectively doing so, and the forecast widespread deployment of PV should eventually lead to aesthetic designs for PV at a reasonable price.

Phases of Market Growth

Putting together all of the economic drivers, noneconomic forces, and potential for governments and industry to stimulate increased deployment of distributed solar energy presents a vision of a radical change in the global energy industry. Such a change from today's fossil-fuel-dependent world to one of distributed-energy generation will proceed in stages, each of which will present its own obstacles but also provides the resources and tools for addressing those of subsequent phases. Many of the perceived constraints to the widespread deployment of distributed PV—including systemwide intermittency, costs, and technical limits—do not account for precisely when these constraints might be binding and what resources would be available to address those constraints at that time. Table 10.1 shows how three main phases in this transformation will occur and the implications of each on various relevant sectors.

In the *rapid-growth phase,* which will likely occur through the year 2020, the market for PV will increase dramatically, on average 30 percent per year to some forty times its current size. Manufacturers will continue to increase the scale of production, and various methods of production for cheaper and more efficient cells that have been in development for many years will ramp up. Initially, homeowners, businesses, and utilities will attempt to find the most effective and economic method of installing PV but will primarily install extensions of the existing methods of rooftop retrofit systems for end users and additional centralized PV and thermal systems for utilities. While the PV industry will still be small relative to the entire energy infrastructure, the early movers will establish themselves in the PV production and supply chain throughout this period and become the major companies of the future PV industry.

In the *displacement phase,* roughly expected between 2020 and 2040, the rate of growth in the industry will slow as the installed base of PV systems becomes a larger and larger proportion of the total new electricity demand. Limits to growth will appear in some markets, but other markets (including many in the developing world) will still be opening up as they reach a point of cost-effectiveness. Over this period, additional methods of storing daytime energy will need to be deployed at either the utility level or by end users to smooth out intermittency issues. Built-in PV at the time of construction will become standard, helping to continue the drop in PV system costs, and utilities will begin to realize that managing a distributed PV electricity system is preferable to losing customers to independent PV/fuel-cell configurations on homes or businesses.

The *dominant phase* after 2040 will see PV as the preferred energy choice for a large majority of locations and applications. New materials or technologies for harnessing PV might need to be employed to continue the trend of falling costs, but by this point the cost of distributed PV will already be clearly economic against centralized sources of electricity. Industrial users will look to adopt PV, quite probably in combination

Table 10.1
PV traces a typical technology S curve as it moves through phases of development with each phase having specific impacts on various sectors.

	Rapid-Growth Phase	Displacement Phase	Dominant Phase
	2005 to 2020	2020 to 2040	2040 and beyond
Time frame cost of installed PV ($ per watt)	**$6.00 → $2.50**	**$2.50 → $1.60**	**$1.60 → $1.00**
Annual PV market (peak GW)	**1.3 GW → 60 GW**	**60 GW → 500 GW**	**500 GW +**
percentage of annual growth in electricity demand, derated	*< 1% → 27%*	*27% → > 100%*	*> 100%*
Global market revenue	**$10 B → $233 B**	**$233 B → $1.3 T**	**$1.3 T +++**
Cumulative installed PV (peak GW)	**5 GW → 237 GW**	**237 GW → 6,900 GW**	**6,900 GW +**
percentage of total electricity supply, derated	*< .05% → 1.8%*	*1.8% → 37%*	*> 37%*
Utility PV efforts	• Trial centralized systems for peak shaving	• Increased incentives to users for adoption	• Transition into energy service providers
Distributed PV users	• Commercial users • Some residential	• Residential users • Solar Home Systems in the developing world	• Industrial users
Key geographic markets	• High-sun and expensive grid electricity	• Medium-sun and grid electricity • Developing world	• Most industrial and developing world applications
Industry development	• Primarily retrofits • Grid-tied	• Built-in PV on homes and businesses standard	• On-site PV integrated with batteries or fuel cells
New technology	• None required • New types of PV system financing developed	• Additional storage required to address intermittency	• Moving beyond silicon cells possibly required

with large fuel-cell installations to grade and smooth power needs. From here, the advantages of modular PV will allow people in developing nations to grow their power supply in line with their income and ability to pay for it. For nearly every nation of the world, the trend toward energy insecurity will have been reversed, and fossil fuels in transportation and heating applications will be rapidly replaced by renewable electricity applications, allowing for continued growth for the PV industry beyond just the conventional electricity market.

In reality, the projected market-growth rates implied in this analysis are by no means guaranteed, and the timeframes for each phase of the industrial transformation are rough approximations. Variances in the actual growth rates will alter the specific timing but will not likely alter the course of the transitions contemplated. However, studies by Stan Bull of the U.S. National Renewable Energy Laboratory and industry veteran Don Aitkin show that any hope of mitigating the worst of the potential climate-change damage from carbon emissions will rely on a combination of PV deployment at these projected growth rates along with significantly improved global energy efficiency and a dramatic rollout of all the other forms of renewable energy—including wind, biomass, and geothermal energy.[3]

Risks to Growth

Many people harbor skepticism that such a widespread industrial transformation can occur, arising from decades of disappointing experiences or from the kind of rational skepticism that should be applied to all claims about the future, particularly those with so much potential to disrupt the existing order. These conceptual obstacles often come in the form of a hesitation or subconscious question that arises as an appeal for new information or perspectives. Information gained in this book can be used to provide answers to some of these compelling questions surrounding solar energy and PV.

Why Now? What's Different This Time?

Many people, particularly those who remember the first wave of interest in commercializing solar-energy technologies in the 1970s, are reluctant to believe arguments that we are at the threshold of a new solar age.

Much of this book has been devoted to explaining the unprecedented shifts toward cost parity made by solar technology in recent years that are creating entirely new opportunities for solar energy today and in the near future.

The reason that this transition to PV is happening now is that for the first time in history photovoltaic electricity has become cost-effective in a number of large grid-tied markets—specifically, for hundreds of thousands of Japanese households, among others. While PV has long enjoyed steady growth in off-grid applications in both the industrialized and developing worlds, these markets have remained limited due to their small size or a lack of the financing to bring down consumers' monthly cost to competitive levels. Never before has solar energy been cost-effective in an industrialized market like Japan with sufficient financing, size, and industrial capability to drive steady cost improvements and market growth. This unique situation in the history of PV technology creates the opportunity to roll out from an established base to progressively more and larger markets in a hierarchical fashion, as discussed in chapter 7. This market development primes the PV industry in such a way that global rollout becomes a question of *when*—a matter of time, volume, and learning.

While Japan's high electricity prices provide a natural motivation to develop solar-energy solutions, this motivation is almost completely offset by Japan's mediocre insolation. Only a strong legislative commitment to develop the PV industry gave the Japanese solar industry the impetus it needed to achieve true cost-competitiveness. Soon markets with slightly lower electricity prices but significantly better solar resources, such as southern Europe and the southwestern United States, will also be able to achieve solar cost equivalence with grid power—even without subsidy support. This important difference will open up the PV market to several hundred million potential adopters within a decade, adding momentum to the transition.

If PV Is Cost Effective, Why Isn't It More Widely Used?

As discussed in previous chapters, achieving growth in any industry depends on more than availability and cost-effectiveness. Growth also requires the development of markets and businesses to deliver the solutions. In the case of solar energy, the supply chain requires manufacturing capability, distributors, integrators, and installers. Market development also requires financing, rationalized building codes, interconnection agreements, and certification and training programs. The growth of the PV market requires that people—consumers, architects, builders, installers, services, and utility executives—all become comfortable with PV technology. All of these factors are in their early stages for PV. Companies are just beginning to discover, for example, that PV is not just clean but also good for the bottom line.

Like any other technology, PV will ultimately run up against physical limits. In the case of PV, these limits are set by cell efficiency, insolation, and the area of rooftops, building exteriors, parking lots, powerline rights-of-way, landfills, brownfields, and other surfaces that can be realistically covered with PV panels. At the present time, however, PV is far from hitting such limits, and photovoltaics will be generating truly vast amounts of electricity long before these limits become constraining in a way that retards overall market growth. The factor that will hold PV growth back in the next few years is a temporary bottleneck in manufacturing capacity for silicon ingots and PV cells and modules. While PV manufacture is less capital intensive than building nuclear or hydroelectric plants, it still requires large amounts of time, money, and machinery. Since it takes time to deploy these capital assets, factory capacity will therefore remain the primary limiting factor on PV market growth in the years ahead.

Are Today's Large Energy Providers Really Going to Let This Happen?
There is a pervasive belief that machinations by large businesses, wealthy individuals, and the government officials that they finance dictate the world we live in, leaving outsiders and upstarts little latitude for affecting the momentum of nations and economies. Unlikely conspiracy theories aside, many people and corporations in the traditional energy field, rightly or wrongly, do perceive renewable or distributed energy as a threat. However, the underlying learning rate for PV technology cannot be easily slowed by any current set of actors. During the 1980s and 1990s, when the largest PV manufacturers in the world were oil companies, including Mobil, BP, and Shell, these corporate giants were limited by both a conflict of interest between protecting their traditional energy operations and promoting PV as well as a technology that was simply

not yet cheap enough to be commercialized without substantial government support.

Today, both conditions are different. The growth in the PV industry over the last decade occurred in Japanese and German microelectronics manufacturers. In the United States, General Electric has entered the market by acquiring AstroPower, until recently the largest U.S. solar-cell manufacturer, and is creating multimillion-dollar research centers in New York and Munich for further development of renewable-energy technologies. These companies do not share the internal conflicts of the existing energy industry players and wield substantial wealth and political clout.

Further, the global companies that comprise the existing energy infrastructure are finding that it is increasingly difficult to influence all of the markets where solar power is gaining momentum. In the 1980s, when the federal government of the United States was the dominant and driving force in PV development, it was relatively easy for the existing fossil-fuel and utility industries to bring pressure to bear on key politicians to stop or slow potential developments.

Today, the governments that are driving these changes include those of states in the American Southwest and Northeast, Japan, and parts of Europe. None of these places harbors significant oil industries, and most have constituencies that are loudly voicing their preference for renewable energy. It is becoming increasingly difficult for traditional energy providers to combat both democratic and corporate efforts in so many jurisdictions.

Is There Enough Solar Energy to Meet Our Needs?

According to the U.S. Department of Energy's Energy Efficiency and Renewable Energy (EERE) office, the amount of sunlight that falls on the earth every day is equivalent to the total energy that is used by the earth's current population in twentyseven years.[4] Not even a large percentage of all of that energy is available to use, but EERE has calculated the total area required to meet U.S. electricity needs using today's PV technology. Even at today's efficiency of PV cells, the land required would be 10 million acres, or 0.4 percent of the total land area of the United States.[5] This would be only 7 percent of the area currently covered by cities and residences, many of which would provide viable locations for integrating PV systems and are unavailable to centralized electricity generators. Comparing PV to other renewable technologies including wind, biomass, and concentrating thermal power suggests that PV requires less land area than any of these technologies for an equivalent amount of energy.

Shouldn't We Focus on More Immediate Alternatives?

In 2004, new solar-energy installations totaled 1.0 GW out of 140 GW of generation installed globally that year—about 0.7 percent, but less than 0.5 percent when derated to compare it to actual electricity generated by other methods. It could seem more productive to focus efforts on other energy solutions (such as wind power, nuclear power, and perhaps clean coal), given that PV's market share is still so small. Despite appearances, this will not happen for two fundamental economic reasons.

First, the ongoing shift toward solar energy will have much greater impact than its present market penetration indicates. Even if the solar industry experiences somewhat lower growth rates over the next decade than it has over the last decade, solar energy could still be accounting for half or more of all electric-generation capacity installed annually fifteen to twenty years from now—at dramatically lower cost to customers than they will be able to get from their local utility providers. Long before the world gets to that point, however, solar energy will recharge people's expectations for the future as they increasingly appreciate its potential to transform energy industry economics. And as more and more people begin to understand the inevitability of the shift to solar, they will be less inclined to bet on marginal or deteriorating solutions. Long before solar power becomes dominant, it will routinely factor into decisions to buy, build, and finance other potential energy solutions, both traditional and renewable. Many large energy projects (such as dams, nuclear power plants, and liquefied natural-gas ports) have construction cycles of years or even decades, which will limit the demand for such investments as the awareness of cost-effective solar electricity continues to rise.

Second, it would be unwise to forgo the enormous wealth effects of accelerating the change to the solar economy. It is simply a smart use of resources to promote these technologies. Unlike discovery of new fossil fuels, discovery of new solar knowhow can be replicated around the world rapidly with little additional cost. While some of the wealth effects to be expected from an increasingly solar economy are nonmeasurable, they will clearly include economic and social growth similar to what resulted from harnessing fossil fuels in the eighteenth and nineteenth centuries and again from the development of the electricity grid in the twentieth century. Because the next paradigm shift—the move to solar energy—will extend to many more people than the current energy infrastructure, it will contribute even more broadly to human prosperity. Benefits to the industrialized world will include increased energy security and stability, cleaner air and water, cheaper electricity, more jobs, and a truly sustainable infrastructure. In the developing world, photovoltaics will do much to help billions of people help themselves out of poverty.

So that is the story of PV, its potential, and the exciting and radical transformation in global energy over the next few decades. Throughout, an effort has been made to be conservative in assumptions and projections, though the conclusions point to changes in the global economy that are no less than revolutionary. In reality, the coming changes will likely exceed not only most current expert projections but our ability to foresee, as many similar technological revolutions throughout history have shown. A mere 150 years ago, not even the optimists anticipated the changes that would soon sweep in rapid succession over the world. Automobiles, telephones, electric power, airplanes, nuclear weapons, plastics, computers, and genetic engineering that have remade our world again and again. An honest assessment of history suggests that rapid technological and industrial revolution is not an aberration; it is the most common outcome.

One of these rapid technological turnovers, the *first silicon revolution*, is particularly relevant to solar energy. Over the last thirty years, the first silicon revolution astonished the world as computers, e-mail, digital music and video, and the cellular telephone went from nonexistent to ubiquitous. In a blindingly short span of time, they emerged to change the world. Each of these technologies, similar in form to photovoltaics, married the power of miniaturization with economies of scale in production. Each offered devices that people wanted at prices that dropped rapidly. In each case, it took intense technical sophistication to begin the cycle, but once the cycle started, it transformed the world in just a few years. In the first silicon revolution, those who were able to anticipate the trend and capitalize on it have become some of today's wealthiest people.

The world currently stands at the beginning of the *second silicon revolution*. It will prove equally profitable—for those who earliest recognize it and take advantage of it. There will be difficulties, obstacles, and challenges in deploying the vast amount of capital and resources required to maintain and grow the standard of living for the world's population, and the coordinated efforts of businesses and governments could help mitigate its risks and accelerate its success. However, moving from a world critically dependent on fossil fuels and a centralized energy-distribution infrastructure toward a renewable world dominated by locally generated solar energy is ultimately unstoppable; the outcome inevitable. Economics and self-interest will catalyze the growth of a solar-energy-based society and provide energy that is not only clean but more affordable and widely available throughout the world. As a result, we will no longer have to make tradeoffs between prosperity and the environment as a result of energy scarcity constraints. Access to cheap, clean, modern energy will spread around the world to become as universal a possession as sunshine itself.

Appendix: Energy and Electricity Measurements

Measurements

The list below shows the prefix for magnitudes of units, whether discussing metrics of energy (Btu or joules) or electricity (Watts or watt-hours):

Kilo thousands (1,000)
Mega millions (1,000,000)
Giga billions (1,000,000,000)
Tera trillions (1,000,000,000,000)
Peta quadrillions (1,000,000,000,000,000)
Exa quintillion (1,000,000,000,000,000,000)

Energy

Measurements of energy used or generated are usually stated in either British Thermal Units (in the United States and occasionally the United Kingdom) or joules (the rest of the world). Standardized energy measurements describe the capacity for each energy source to generate heat (Btu) or perform work (joule). They can be converted back and forth by

1 Btu equals approximately 1060 joules.

As figure 2.2 shows, the United States in 2004 consumed 99.8 quadrillion Btus, which would be the equivalent of 105.6 exajoules. Globally, roughly 400 exajoules, or 378 quadrillion Btus, of modern energy are used each year.

Electricity

In electricity, the concepts of an electricity generator's *peak capacity* and *electricity generated* are important to distinguish for many discussions in this book.

- *Peak capacity* measures the maximum amount of electricity at any moment that can be provided by a given generator, and it is usually measured in kilowatts (kW), megawatts (MW, thousands of kilowatts), or gigawatts (GW, millions of kilowatts).
- *Electricity generated* is the amount of electricity actually produced by a generator and is usually measured in terms of hours at the peak capacity, or kilowatt-hours (kWh), megawatt-hours (MWh), gigawatt-hours (GWh), or terawatt-hours (TWh). For comparison, kWh can be easily converted to other measurements of energy:

1 kWh equals 3.4 Btu
1 kWh equals 3600 joules.

Switching between measurements of peak capacity and electricity generated is easily done by multiplying the peak capacity by the number of hours it is used. Therefore, a 250 MW peak generator that is used at its peak capacity for 2000 hours would generate 500 GWh of electricity (or 1.7 billion Btus).

Notes

Chapter 1

1. Thomas Malthus, "An Essay on the Principle of Population," Electronic Scholarly Publishing Project, 1998 (originally published in 1798). Available at <http://www.esp.org/books/malthus/population/malthus.pdf>.

2. Energy Information Administration (EIA), *International Outlook 2004* (Washington, D.C.: U.S.Government Printing Office), p. 1. Available at <http://www.eia.doe.gov/oiaf/archive/ieo04/pdf/0484(2004).pdf>.

3. International Energy Agency (IEA), *Key World Energy Statistics 2005* (Paris: IEA, 2005), p. 6. Available at <http://www.iea.org/bookshop/add.aspx?id=144>.

4. IEA, *Key Energy Statistics 2005*, pp. 6, 28.

5. National Academy of Engineering, "Greatest Engineering Achievements of the 20th Century," retrieved March 30, 2004, from <http://www.greatachievements.org> (adapted from George Constable and Bob Somerville, *A Century of Innovation: Twenty Engineering Achievements That Transformed Our Lives* [Washington, D.C.: Joseph Henry Press, 2003]).

6. EIA, "Electric Power Monthly September 2005 with Data for June 2005," (Washington, D.C.: U.S.Government Printing Office), pp. 108–109. Available at <http://tonto.eia.doe.gov/FTPROOT/electricity/epm/02260509.pdf>.

7. IEA, *Key World Energy Statistics 2005*, p. 43.

8. *Cost of delivery* here includes the cost of both transmission and distribution in aggregate.

9. Xcel Energy, "What Does It Cost to Produce Your Electricity?" (Minneapolis: Xcel Energy, 2004). Available at <http://www.xcelenergy.com/docs/corpcomm/generationtransmissiondistributioncostsonbill.pdf>.

10. Denise Warkentin, *Electric Power Industry in Nontechnical Language* (Tulsa, OK: PennWell, 1998), p. 112.

11. American Wind Energy Association (AWEA), "The Wind Energy Tutorial: Wind Energy Costs" (Washington, DC: AWEA, 2005), retrieved October 10, 2005, from <http://www.awea.org/faq/tutorial/wwt_costs.html>.

12. World Wind Energy Association (WWEA), “Press Release: Worldwide Wind Energy Capacity at 47.616 MW, 8.321 MW added in 2004,” March 7 2005. Available at <http://www.wwindea.org/default.htm>.

13. World Nuclear Association (WNA), “Nuclear Power in the World Today,” (London: WNA, 2004), retrieved January 4, 2005, from <http://www.world-nuclear.org/info/inf01.htm>.

14. Pietro S. Nivola, “The Political Economy of Nuclear Energy in the United States,” *Policy Brief #138*, September 2004 (Washington, D.C.: The Brookings Institution, 2004). Available at <http://www.brookings.edu/comm/policybriefs/pb138.htm>.

15. Bruce Kennedy, “China’s Three Gorges Dam,” retrieved January 4, 2005, from <http://www.cnn.com/SPECIALS/1999/ china.50/asian.superpower/three.gorges/>.

16. For more information on this topic, see World Commission on Dams at <http://www.dams.org>.

17. This assertion is based on discussions with industry participants who feel that large systems (over 20 kW) can be installed at a 15 to 25 percent discount over small systems (less than 5 kW). The cost of capital for small systems is cheaper than that available for large systems by an offsetting amount.

18. Solarbuzz Inc., *Market Buzz 2005: Annual World PV Market Review* (San Francisco: Solarbuzz Inc., 2005), pp. 14, 100.

19. Solarbuzz, *Market Buzz 2005*, p. 14.

20. Jack Casazza and Frank Delea, *Understanding Electric Power Systems: An Overview of the Technology and the Marketplace* (Hoboken, NJ: Wiley, 2003), p. 67.

21. The generator market-share percentages estimated here are based on an estimated 1.2 GW of annual installation of PV in 2005, versus global electricity generator installation of 120 to 140 GW in the same year coming from extrapolating various data sources. The estimates of percentage of electricity generated assume that 5 GW of cumulative global PV installations used an average of 1,000 AC hours per year, divided by the EIA estimates (available at <http://www.eia.doe.gov/aer>), of 16,661 thousand gigawatt hours (TWh) per year generated globally.

22. Amory Lovins et al., *Small Is Profitable: The Hidden Economic Benefits of Making Electrical Resources the Right Size* (Snowmass, CO: Rocky Mountain Institute, 2002), p. 118.

23. Solarbuzz, *Market Buzz 2005*, p. 14.

24. Anne Kreutzmann, “Rush to Trading Floor: PV Companies Flock towards the Stock Market,” *Photon International* (October 2005): 52.

25. European Renewable Energy Council (EREC), “Renewable Energy Scenario to 2040: Half of the Global Energy Supply from Renewables in 2040” (Brussels: EREC, June 2004). Available at <http://www.erec-renewables.org/documents/targets_2040/EREC_Scenario 2040.pdf>.

Chapter 2

1. For ease of explanation, the Protista, Fungi, and Monera kingdoms of animals are not included here, but their inclusion would not alter the point.

2. Niles Eldridge, "The Sixth Extinction" (Washington, DC: American Institute of Biological Sciences, 2001), retrieved July 19, 2004 from <http://www.actionbioscience.org/newfrontiers/eldredge2.html#Primer>.

3. Janet Larsen, "Forest Cover Shrinking" (Washington, DC: Earth Policy Institute, 2002), retrieved July 29, 2004, from <http://www.earth-policy.org/Indicators/indicator4.htm>.

4. Clive Ponting, *A Green History of the World: The Environment and Collapse of Great Civilizations* (New York: Penguin Books, 1991), p. 24.

5. Ponting, *A Green History of the World*, p. 39.

6. Jared Diamond, *Guns, Germs, and Steel: The Fates of Human Societies* (New York: Norton, 1999).

7. Joseph A. Tainter, *The Collapse of Complex Societies* (Cambridge: Cambridge University Press, 2003), p. 50.

8. Ponting, *A Green History of the World*, p. 71.

9. Ponting, *A Green History of the World*, p. 72.

10. Ponting, *A Green History of the World*, p. 72.

11. Ponting, *A Green History of the World*, p. 72.

12. Jeremy Rifkin, *The Hydrogen Economy* (New York: Tarcher/Putnam, 2002), p. 59.

13. Gavin Menzies, *1421: The Year China Discovered America* (New York: Perennial, 2004), p. 58.

14. William McNeill, *Plagues and Peoples* (New York: Doubleday/Anchor Books, 1976), p. 147.

15. Ponting, *A Green History of the World*, p. 281.

16. Vaclav Smil, *Energy in World History* (Boulder, CO: Westview Press, 1994), p. 161.

17. Rifkin, *The Hydrogen Economy*, p. 68.

18. Rifkin, *The Hydrogen Economy*, p. 65.

19. Denise Warkentin, *Electric Power Industry in Nontechnical Language* (Tulsa, OK: PennWell, 1998), p. 31.

20. Warkentin, *Electric Power Industry*, p. 31.

21. Jack Casazza and Frank Delea, *Understanding Electric Power Systems: An Overview of the Technology and the Marketplace* (Hoboken, NJ: Wiley, 2003), p. 1.

22. Warkentin, *Electric Power Industry*, pp. 2, 37.

23. Warkentin, *Electric Power Industry*, p. 39.

24. Warkentin, *Electric Power Industry*, p. 62.

25. Vijay V. Vaitheeswaran, *Power to the People: How the Coming Energy Revolution Will Transform an Industry, Change Our Lives, and Maybe Even Save the Planet* (New York: Farrar, Straus and Giroux, 2003), p. 31.

26. Casazza and Delea, *Understanding Electric Power Systems*, p. 19.

27. Casazza and Delea, *Understanding Electric Power Systems*, p. 5.

28. Daniel Yergin, *The Prize: The Epic Quest for Oil, Money, and Power* (New York: Touchstone/Simon & Schuster, 1992), p. 14.

29. Yergin, *The Prize*, p. 28.

30. "Automobile Industry: Industry History," Columbia Encyclopedia retrieved January 6, 2005, from <http://www.encyclopedia.com/html/section/automobind_IndustryHistory.asp>.

31. Yergin, *The Prize*, p. 13.

32. Yergin, *The Prize*, p. 13.

33. Yergin, *The Prize*, p. 13.

34. Siamack Shojai (Ed.), *The New Global Oil Market: Understanding Energy Issues in the World Economy* (Westport, CT: Praeger, 1995), p. 85.

35. Energy Information Administration (EIA), "Country Analysis Briefs: OPEC Brief," November 8, 2005 (Washington, DC: U.S. Department of Energy, 2005), retrieved January 6, 2005, from <http://www.eia.doe.gov/emeu/cabs/opec.html>.

36. Warkentin, *Electric Power Industry*, p. 63.

37. Warkentin, *Electric Power Industry*, p. 63.

38. Natural Gas Supply Association, "Overview of Natural Gas: History," retrieved January 6, 2006, from <http://www.naturalgas.org/overview/history.asp>.

39. Natural Gas Supply Association, "Overview of Natural Gas: History."

40. "Global 500: World's Largest Corporations," July 25, 2005, *Fortune*, p. 119. Also available at <http://www.fortune.com/fortune/global500/fulllist/0,24394,1,00.html>.

41. The revenues of Fortune 500 companies are listed at <http:// www.fortune.com/fortune/global500/fulllist/0,24394,1,00.html>; global gross domestic product (GDP) ($55.5 trillion) from Central Intelligence Agency, "The World Factbook," November 1, 2005. Available at <http://www.cia.gov/cia/publications/factbook/geos/xx.html>.

42. Conversation with John Holdren, October 18, 2005.

43. International Energy Agency (IEA), *Key World Energy Statistics 2005* (Paris: IEA, 2005). Available at <http://www.iea.org/bookshop/add.aspx?id=144>.

44. EIA, "U.S. Primary Energy Consumption by Source and Sector, 2004," August, 15, 2005, from *Annual Energy Outlook 2004*, Report No. DOE/EIA-0384(2004) (Washington, D.C.: U.S.Government Printing Office, 2004). Available at <http://www.eia.doe.gov/emeu/aer/pecss_diagram.html>.

45. IEA, *Key World Energy Statistics 2005*.

46. Howard Geller, *Energy Revolution: Policies for a Sustainable Future* (Washington, DC: Island Press, 2003), p. 131.

47. International Energy Agency (IEA), *Toward a Sustainable Energy Future* (Paris: OECD, 2001). Available at <http://www.iea.org/textbase/nppdf/free/2000/future2001.pdf>.

48. Jose Goldberg (ed.), "Overview," in *World Energy Assessment: Energy and the Challenge of Sustainability* (New York: United Nations Development Program [UNDP], 2000). Available at <http://www.undp.org/seed/eap/activities/wea/index.html>.

49. IEA, *World Energy Outlook 2002* (Paris: OECD, 2001), chapter 13 (Energy and Poverty), p. 13.

50. IEA, *Key World Energy Statistics 2005.*

51. EIA, *Annual Energy Outlook 2004*, Report No. DOE/EIA-0384(2004) (Washington, D.C.: U.S.Government Printing Office, 2004). Available at <http://www.eia.doe.gov/aer>.

52. Platts, "Global Power Plant Additions Reach Unprecedented Levels, According to Platts," April 5, 2004. Available at <http://www.platts.com/Resources/Press Room/2004/040504.xml?S=printer>; IEA, *Key World Energy Statistics 2005.*

Chapter 3

1. World Bank, *World Development Report 2003* (New York: World Bank, 2003), p. 2.

2. United Nations Population Division (UNPD), *World Population Prospects: The 2002 Revision* (New York: United Nations, 2002). Available at <http://www.un.org/esa/population/publications/wpp2002/WPP2002-HIGHLIGHTSrev1.PDF>.

3. The Worldwatch Institute, *Vital Sign 2005: The Trends That Are Shaping our Future* (New York: W.W. Norton & Company, 2005), p. 65.

4. Lester R. Brown, *Eco-Economy: Building an Economy for the Earth* (New York: Norton, 2001).

5. RAND Institute, "International Family Planning: A Success Story So Far," Document No: RB-5022 (Santa Monica, CA: RAND Institute, 1998), retrieved July 20, 2004, from <http://www.rand.org/publications/RB/RB5022/index.html>.

6. United States Census Bureau, "Global Population Profile: 2002."

7. International Energy Association (IEA), *World Energy Outlook 2002* (Paris: OECD, 2001), chapter 13 (Energy and Poverty), p. 13.

8. United Nations Development Program (UNDP), *Energy for Sustainable Development: A Policy Agenda,* Thomas B. Johansson and Jose Goldberg (eds.), (New York: UNDP, 2002), p. 12.

9. Michael T. Klare, *Resource Wars: The New Landscape of Global Conflict* (New York: Holt, 2001), appendix.

10. David Urbinato, "London's Historic "Pea-Soupers," retrieved March 16, 2004, from <http://www.epa.gov/history/topics/perspect/london.htm>.

11. "Science and Health: Deadly Smog," NOW with Bill Moyers, January 17, 2003, retrieved March 16, 2004, from <http://www.pbs.org/now/science/smog.html>.

12. Vijay V. Vaitheeswaran, *Power to the People: How the Coming Energy Revolution Will Transform an Industry, Change Our Lives, and Maybe Even Save the Planet* (New York: Farrar, Straus and Giroux, 2003), p. 179.

13. Vaitheeswaran, *Power to the People*, p. 295.

14. Susmita Dasgupta, Hua Wang, and David Wheeler, "Surviving Success: Policy Reform and the Future of Industrial Pollution in China," (Washington, DC: World Bank, 1997), retrieved January 13, 2005, from <http://www.worldbank.org/nipr/work_paper/survive/>.

15. United Nations Environment Program (UNEP), *The Asian Brown Cloud* (New York: United Nations, 2002).

16. Bruce Sundquist, "Topsoil Loss: Causes, Effects, and Implications: A Global Perspective," retrieved July 28, 2004, from <http://home.alltel.net/bsundquist1/se1.html>.

17. Brown, *Eco-Economy*, p. 50.

18. Brown, *Eco-Economy*, p. 149.

19. Lester R. Brown, *World Food Security Deteriorating* (Washington, DC: Earth Policy Institute, 2004), retrieved January 13, 2005, from <http://www.earth-policy.org/Updates/Update40.htm>.

20. Rainforest Action Network (RAN), "About Rainforests: By the Numbers," retrieved July 28, 2004, from <http://www.ran.org/info_center/about_rainforests.html>.

21. RAN "About Rainforests"

22. Jeremy Lovell, "Fresh Studies Support New Mass Extinction Theory," March 19, 2004, retrieved January 13, 2005, from <http://www.planetark.com/dailynewsstory.cfm/newsid/24346/newsDate/19-Mar-2004/story.htm>.

23. David A. King, "Environment: Climate Change Science: Adapt, Mitigate, or Ignore?" *Science* 303(5655) (January 2004):176–177.

24. Intergovernmental Panel on Climate Change (IPCC), *Climate Change 2001: The Scientific Basis* (Cambridge: Cambridge University Press, 2001). Available at <http://www.grida.no/climate/ipcc_tar/wg1/pdf/WG1_TAR-FRONT.PDF>.

25. Munich Re Group, *Topics Geo: Annual Review, Natural Catastrophes 2004* (Munich: Munich Re Group, 2004). Available at <http://www.munichre.com/publications/302-04321_en.pdf?rdm=66061>.

26. Munich Re Group, *Topics Geo: Annual Review, Natural Catastrophes 2004.*

27. Munich Re Group, *Topics Geo: Annual Review, Natural Catastrophes 2004.*

28. IPCC, *Climate Change 2001.*

29. Geoff Jenkins, "Current Science of Climate Change," presented at the UNEP Industry Consultative Meeting, Paris, France, October 9, 2003. Available at <http://www.uneptie.org/outreach/business/docs/2003presentations/Thur-panel%201-Jenkins%201.pdf>.

30. Energy Information Administration (EIA), *International Energy Outlook 2005*, Report No. DOE/EIA-0484 (Washington, DC: U.S. Government Printing Office, 2005). Available at <http://www.eia.doe.gov/oiaf/ieo/pdf/0484(2005).pdf>.

31. EIA, *International Energy Outlook 2005.*

32. Daniel Yergin, *The Prize: The Epic Quest for Oil, Money, and Power* (New York: Touchstone/Simon & Schuster, 1992).

33. Kenneth S. Deffeyes, *Hubbert's Peak: The Impending World Oil Shortage* (Princeton, NJ: Princeton University Press, 2001), p. 2.

34. Jeff Gerth, "Forecast of Rising Oil Demand Challenges Tired Saudi Fields," *New York Times,* February 25, 2004, Section A, Column 3, Business/Financial Desk, p. 1.

35. Matthew R. Simmons, "Is There an Energy Crisis?" presented at the IPAA, IPAMS, DESK & DERRICK Group lunch October 11, 2004. Available at <http://www.simmonsco-intl.com/files/IPAA-IPAMS-DESK%20&%20DERRICK.pdf>.

36. British Petroleum, *Statistical Review of World Energy* (London: BP, 2003).

37. Paul Roberts, *The End of Oil: On the Edge of a Perilous New World* (Boston, Houghton Mifflin, 2004), p. 58.

38. Robert Ebel, "Russian Reserves and Oil Potential," Paper presented at the Centre for Global Energy Studies meeting, March 15, 2004. Ebel cites an internal study from Yukos, which at the time was one of Russia's leading oil firms.

39. Howard Geller, *Energy Revolution: Policies for a Sustainable Future* (Washington, DC: Island Press, 2003), p. 11.

40. Tim Appenzeller, "The End of Cheap Oil," *National Geographic* (June 2004): 106.

41. EIA, *International Energy Outlook 2005.*

42. Edison Electric Institute (EEI), "Comments on Residential Furnace and Boiler Standards," Docket No. EE-RM/STD-01-350, November 9, 2004, Energy Efficiency and Retail Markets. Available at <http://www.eei.org/about_EEI/advocacy_activities/U.S._Department_of_Energy/ EEI_Furnace-Boiler-ANOPR_041109.pdf>.

43. Richard Wilson, "Combating Terrorism an Event Tree Approach," March 19, 2003. Available at <http://www.whitehouse.gov/omb/inforeg/2003report/356.pdf>.

44. Julian Darley, "Why Natural Gas Is Not an Alternative," paper presented at the Second World Renewable Energy Forum, Bonn, Germany, May 31, 2004.

45. David R. Francis, "The Escaping Price of Natural Gas," February 19, 2004, *The Christian Science Monitor*, retrieved on August 4, 2004, from <http://www.csmonitor.com/2004/0219/p11s02-usec.html>.

46. "Exxon Says North America Gas Production Has Peaked," June 21, 2005. Available at <http://reuters.com>.

47. EEI, "Natural Gas," retrieved January 13, 2006, from <http://www.eei.org/industry_issues/energy_infrastructure/natural_gas>.

48. Union of Concerned Scientists (UCS), "Renewable Energy Can Help Ease Natural Gas Crunch," August 26, 2005, clean energy fact sheet. Available at <http://www.ucsusa.org/clean_energy/renewable_energy/page.cfm?pageID=1370>.

49. Hans-Holger Rogner (ed), "Energy Resources" in *World Energy Assessment: Energy and the Challenge of Sustainability* (New York: United Nations Development Program (UNDP), 2000). Available at <http://www. undp.org/seed/eap/activities/wea/index.html>.

50. Jeremy Rifkin, *The Hydrogen Economy* (New York: Tarcher/Putnam, 2002), p. 130.

51. "Republic of South Africa Special Edition," *Ecoal*, September 1999. Available at <http://www.worldcoal.org/assets_cm/files/PDF/ecoal_1999_september_rsa_special_1.pdf>.

52. Ken Silverstein, "Coal Liquefaction Plants Spark Hope," November 1, 2004, *Daily IssueAlert*, retrieved January 13, 2005, from <http://www.utilipoint.com/issuealert/article.asp?id=2314>.

53. Mai Tian, "First Coal Liquefaction Centre Set Up in Shanghai," *China Daily*, March 12, 2004, retrieved January 13, 2005, from <http://www.chinadaily.com.cn/english/doc/2004-03/12/content_314118.htm>.

54. British Petroleum, *Statistical Review of World Energy*, pp. 4, 20.

55. British Petroleum, *Statistical Review of World Energy*, pp. 4, 20.

56. Geller, *Energy Revolution*, p. 11.

57. Geller, *Energy Revolution*, p. 11.

58. Youssef Ibrahim, "The World Has Lost Iraq's Oil," *USA Today*, October 5, 2004, retrieved January 13, 2005, from <http://www.usatoday.com/news/opinion/editorials/2004-10-05-iraqi-oil_x.htm>.

59. EIA, "World Oil Transit Chokepoints" (Washington, DC: U.S. Government Printing Office, 2004), retrieved July 14, 2004, from <http:// www.eia.doe.gov/emeu/cabs/choke.html>.

60. Vaitheeswaran, *Power to the People*, p. 111; Rifkin, *The Hydrogen Economy*, p. 35; Roberts, *The End of Oil*, p. 183.

61. Bhushan Bahree and Gregory White, "World Oil Supply Faces Stress in Months Ahead," *Wall Street Journal*, July 14, 2004, p. A2.

62. Jack Casazza and Frank Delea, *Understanding Electric Power Systems: An Overview of the Technology and the Marketplace* (Hoboken, NJ: Wiley, 2003), pp. 9–10.

63. Masayki Yajima, *Deregulatory Reforms of Electricity Supply Industry* (New York: Quorum Books, 1997), p. 21.

64. Casazza and Delea, *Understanding Electric Power Systems*.

65. James L. Sweeney, "The California Electricity Crisis: Lessons for the Future," *The Bridge* 32, n.2 (2002), retrieved September 1, 2004, from <http://www.nae.edu/nae/nawhome.nsf/weblinks/MKEZ-5B7JJ2?OpenDocument>.

66. EIA, *International Energy Outlook 2004* (Washington, D.C.: U.S. Government Printing Office). Available at <http://www.eia.doe.gov/oiaf/archive/ieo04/pdf/0484(2004).pdf>.

67. "Europe's Power Struggle," *The Economist*, July 3, 2004, 49–50.

Chapter 4

1. Wim C. Turkenburg (ed), "Renewable Energy Technology," in *World Energy Assessment: Energy and the Challenge of Sustainability* (New York: United Nations Development Program [UNDP], 2000). Available at <http://www.undp.org/seed/eap/activities/wea/index.html>.

2. International Energy Agency (IEA), *Renewables Information* (Paris: IEA, 2003), p. 11; World Commission on Dams, *Dams and Development: A New Framework for Decision-Making* (London: Earthscan, 2000), p. 14.

3. Glenn Switkes, "Molten Rivers: The Aluminum and Hydroelectric Dam Connection" (São Luís, Brasil: International Strategic Roundtable on the Aluminum Industry, 2003). Available at <http://www.irn.org/programs/aluminium/aluminumreport2003.pdf>.

4. World Commission on Dams, *Dams and Development: A New Framework for Decision-Making* (London: Earthscan, 2000), p. 104.

5. Patrick McCully, *Silenced Rivers: The Ecology and Politics of Large Dams* (London: Zed Books, 2001).

6. Department of Energy (DOE), "Hydropower Resource Potential," 2005, Energy Efficiency and Renewable Energy website, available at <http://www.eere.energy.gov/windandhydro/hydro_potential.html>.

7. "Damming Evidence," *The Economist*, July 19, 2003, pp. 9–11.

8. Stephen Hilgartner, Richard C. Bell, and Rory O'Connor, *Nukespeak: The Selling of Nuclear Technology in America* (New York: Penguin, 1982), p. 44.

9. World Nuclear Association (WNA), "Nuclear Power in the World Today," March 2004, retrieved July 14, 2004, from <http://www.world-nuclear.org/info/inf01.htm>.

10. Pietro S. Nivola, "The Political Economy of Nuclear Energy in the United States," Policy Brief #138, September 2004, (Washington, D.C.: The Brookings Institution, 2004). Available at <http://www.brookings.edu/comm/policybriefs/pb138.htm>.

11. Pietro S. Nivola, "The Political Economy of Nuclear Energy in the United States.

12. Pietro S. Nivola, "The Political Economy of Nuclear Energy in the United States.

13. Federation of American Scientists (FAS), "India: Nuclear Weapons" (Washington, DC: FAS, 2002). Available at <http://www.fas.org/nuke/guide/india/nuke/>.

14. Richard Tomlinson, "AREVA: The Queen of Nukes," *Fortune,* May 17, 2004, p. 55.

15. "Germany Starts Nuclear Energy Phase-Out," *Deutsche Welle*, November 14, 2003. Available at <http://www.dw-world.de/dw/article/0,,1029748,00.html>.

16. IEA, *Renewables Information*, pp. 11, 15.

17. John Deutch, Ernest J. Moniz, et al., *The Future of Nuclear Power: An MIT Interdisciplinary Study* (Cambridge, MA: MIT, 2003). Available at <http://web.mit.edu/nuclearpower/>.

18. Amory B. Lovins and L. Hunter Lovins, "The Nuclear Option Revisited," *Los Angeles Times*, July 8, 2001. Available at <http://www.rmi.org/images/other/Energy/E01-19_NuclearOption.pdf>.

19. Norman Myers and Jennifer Kent, *Perverse Subsidies: How Tax Dollars Can Undercut the Environment and the Economy* (Washington, DC: Island Press, 2001), p. 78.

20. Nuclear Energy Institute (NEI), "NEI Applauds Senate Approval of Amendment to Renew Price-Anderson Act" (Washington, DC: NEI, 2002). Available at <http://www.nei.org/index.asp?catnum=4&catid=414>.

21. Jackie Jones, "Onward and Upward," *Renewable Energy World* (July–August 2004): 60.

22. Global Wind Energy Council (GWEC), "Global Wind Energy Market Will Double to A$25 Billion per Year by 2010," August 17, 2005 (Sydney: GWEC). Available at <http://www.gwec.net/fileadmin/documents/PressReleases/050815 intl_media_release.pdf>.

23. American Wind Energy Association (AWEA), "The Economics of Wind Energy," (Washington, DC: AWEA, 2005). Available at <http://www.awea.org/pubs/factsheets/EconomicsOfWind-Feb2005.pdf>.

24. AWEA, "The Economics of Wind Energy."

25. European Wind Energy Association (EWEA), "Windpower Economics," (Brussels: EWEA, n.d.). Available at <http://www.ewea.org/fileadmin/ewea_documents/documents/publications/factsheets/factsheet_economy2.pdf>.

26. AWEA, "Global Wind Energy Market Report" (Washington, DC: AWEA, 2002). <http://www.awea.org/pubs/documents/globalmarket2003.pdf>.

27. Conversation with Josh Green, Massachusetts Technology Collaborative, August 28, 2004.

28. "Ill Winds," *The Economist*, July 29, 2004, retrieved August 2, 2004, from <www.economist.com>.

29. Marlise Simons, "Where Nelson Triumphed, a Battle Rages over Windmills," *New York Times*, January 10, 2005, p.A4.

30. Simons, "Where Nelson Triumphed."

31. CBS News, "Storm over Mass. Windmill Plan," June 29, 2003. Available at <http://www.cbsnews.com/stories/2003/06/26/sunday/main560595.shtml>.

32. Turkenburg, "Renewable Energy Technology," p. 222.

33. Oak Ridge National Laboratory, "Energy Crops and the Environment." Available at <http://bioenergy.ornl.gov/papers/misc/cropenv.html>.

34. "U.S. versus the World: In Europe, Biodiesel's King. In Brazil, Ethanol Rules," *Soybean Digest*, August 1, 2002. Available at <http://www.cornandsoybeandigest.com/mag/soybean_us_versus_world/>.

35. World Bank, "Geothermal Energy" (New York: World Bank). Available at <http://www.worldbank.org/html/fpd/energy/geothermal/>.

36. G.W. Huttrer, "The Status of World Geothermal Power Generation 1995-2000," *Geothermics* 30, n.1 (2001): 1–28.

37. World Bank, "Geothermal Energy."

38. Karl Gawell, Marshall Reed and P. Michael Wright, "Preliminary Report: Geothermal Energy, the Potential for Clean Power from the Earth" (Washington, DC: Geothermal Energy Association, 1999). Available at <http://www.geo-energy.org/PotentialReport.htm>.

39. Jose Goldberg and Thomas B. Johansson (eds.), *World Energy Assessment: Overview 2004 Update* (New York: UNDP, 2004). Available at <http://www.undp.org/energy/docs/WEAOU_part_IV.pdf>, table 6.

40. International Fusion Energy Organization (IFEO) / International Thermonuclear Experimental Reactor (ITER), "What Is the Development Programme?" 2004. Available at <http://www.iter.org/Developfusion.htm>.

41. Cynthia Tornquist, "Nuclear Fusion Still No Dependable Energy Source," *CNN Online*, April 15, 1997. Available at <http://www.cnn.com/US/9704/05/fusion.confusion/index.html>.

42. "Bouillabaisse Sushi," *The Economist*, February 7, 2004, p. 74.

43. Jesse H. Ausubel, "Decarbonization: The Next One Hundred Years," paper presented at the Fiftieth Anniversary Symposium of the Geology Foundation, Jackson School of Geosciences, University of Texas, Austin, Texas, April 25, 2003. Available at <http://phe.rockefeller.edu/AustinDecarbonization/>.

44. Joseph J. Romm, *The Hype about Hydrogen: Fact and Fiction in the Race to Save the Climate* (Washington, DC: Island Press, 2004).

45. Romm, *The Hype about Hydrogen*, pp. 71, 72.

46. Rosa C. Young, "Advances of Solid Hydrogen Storage Systems," Paper presented at the National Hydrogen Association meeting, March 5, 2003, Ovonic Hydrogen Systems, LLC. Available at <http://www.ovonic-hydrogen.com/home>.

47. Marianne Mintz, et al., "Cost of Some Hydrogen Fuel Infrastructure Options," Paper presented to the Transportation Research Board, January 16, 2002 (Chicago: Argonne National Laboratory). Available at <http://www.transportation.anl.gov/pdfs/AF/224.pdf>.

48. Prius Specifications, available at <http://www.toyota.com/vehicles/2005/prius/specs.html>.

49. Socolow, Robert (ed.), "Fuels Decarbonization and Carbon Sequestration: Report of a Workshop," Mechanical and Aerospace Engineering Department,

Princeton University, 1997. Available at <http://mae.princeton.edu/index.php?app=download&id=91>.

50. Barry C. Lynn, “Hydrogen’s Dirty Secret,” *Mother Jones* (May–June 2003). Available at <http://www.motherjones.com/news/outfront/2003/05/ma_375_01.html>.

Chapter 5

1. Martin Green, *Power to the People: Sunlight to Electricity Using Solar Cells* (Sydney: University of New South Wales Press, 2000), p. 16

2. This section draws on exceptional texts that explore the history of the use of solar energy, including Ken Butti and John Perlin, *A Golden Thread: Twenty-five Hundred Years of Solar Architecture and Technology* (Palo Alto, CA: Cheshire Books, 1980); Frank T. Kryza, *The Power of Light: The Epic Story of Man’s Quest to Harness the Sun* (New York: McGraw-Hill, 2003); and Martin Green, *Power to the People*. Each of these books is worth reading to understand the deeply embedded role that solar energy has played and will continue to play in our world.

3. Butti and Perlin, *A Golden Thread,* p. 3.

4. Kryza, *The Power of Light*, p. 37.

5. Kryza, *The Power of Light*, p. 56.

6. Kryza, *The Power of Light*, pp. 91–106.

7. Butti and Perlin, *A Golden Thread*, p. 74.

8. Kryza, *The Power of Light*, p. 171.

9. Kryza, *The Power of Light*, p. 173.

10. Butti and Perlin, *A Golden Thread*, p. 77.

11. Kryza, *The Power of Light*, p. 113.

12. Kryza, *The Power of Light*, p. 209.

13. Butti and Perlin, *A Golden Thread*, p. 89.

14. Kryza, *The Power of Light*, p. 240

15. Green, *Power to the People,* p. 22.

16. Solarbuzz Inc, *MarketBuzz 2005: Annual World PV Market Review* (San Francisco: Solarbuzz Inc., 2005), p. 15.

17. Solarbuzz, *Market Buzz 2005*, pp. 14, 100.

18. The data used in this section on the current state of the PV industry were generously provided by both Paul Maycock of *PV News*, who has provided valuable industry data for many years, as well as Craig Stevenson of Solarbuzz, whose Web site (<http://www.solarbuzz.com>) and reports on these data are well researched and useful.

19. Paul Maycock, “Market Update: Global PV Production Continues to Increase,” *Renewable Energy World* 8(4) (July–August 2005): 86–99.

20. Maycock, "Market Update," p. 88.

21. Solarbuzz, *Market Buzz 2005*, p 14.

22. Solarbuzz, *Market Buzz 2005*, p. 14.

23. Maycock, "Market Update," p. 88.

24. Kosuke Kurowaka and Kazuhiko Kato, "Power from the Desert," *Renewable Energy World* (January 2003), (London: Earthscan). Available at <http://www.earthscan.co.uk/news/article/mps/UAN/73/v/3/sp/>.

25. Solarbuzz, *Market Buzz 2005*, p. 15.

26. Solarbuzz, *Market Buzz 2005* p. 17.

27. William P. Hirshman, "Small Talk and Big Plans in Japan," *Photon International* (May 2004): 65–67.

28. International Energy Agency (IEA), *Key World Energy Statistics 2005* (Paris: IEA, 2005). Available at <http://www.iea.org/bookshop/add.aspx?id=144>.

29. Maycock, "Market Update," p. 97.

30. Solarbuzz, *Market Buzz 2005*, p. 15.

31. Solarbuzz, *Market Buzz 2005*, p. 14.

32. Solarbuzz, *Market Buzz 2005*, p. 147.

33. Solarbuzz, *Market Buzz 2005*, p. 14.

34. Solarbuzz, *Market Buzz 2005*, p. 15.

35. Maycock, "Market Update," p. 90.

36. Solarbuzz, *Market Buzz 2005*, p. 30

37. Maycock, "Market Update," p. 88.

38. Solarbuzz, *Market Buzz 2005*, p. 14.

39. Maycock, "Market Update," p. 97.

40. Maycock, "Market Update," p. 88.

41. Martin Green, "Third-Generation Solar: Future Photovoltaics," *Renewable Energy World* (July–August 2004): 203.

42. The assumptions built into this model include market growth, experience-curve rates, solar resource for three different locations, and type of financing used to pay for the system. The market-growth rates assumed are declining: 30 percent per year in 2005 to 2020, 25 percent per year in 2021 to 2025, 20 percent per year in 2026 to 2030, and gradually declining to 2 percent between 2031 and 2040 as market saturation begins to occur. The experience-curve rates also are declining over time as the industry grows toward scale production. They begin at 15 percent in 2005 and decline by 2.5 percent each decade (for example, 12.5 percent from 2015 to 2024). The amount of sun is estimated in hours per year and is high (1,900 hours, roughly the amount available in Los Angeles or Phoenix), medium (1,450 hours, Philadelphia), and low (1,100 hours, less than Seattle). These are representative of the many locations that are above the high and below the low levels of sun assumed. Finally, the mortgage assumption for residential PV systems assumes a thirty-year mortgage at 3 percent real rate of interest and a

12 percent loss factor converting DC to AC. The prices are forecast for the U.S. residential market and assume the standard tax deductibility of mortgage interest but no other subsidies, green tags, or other government payments.

43. Solar Electric Power Association, "Policy Statement of the Solar Electric Power Association," November 14, 2002. Available at <http://www.solarelectricpower.org/ewebeditpro/items/O63F3859.pdf>.

Chapter 6

1. Energy Information Administration (EIA), *International Energy Outlook 2005*, Report No. DOE/EIA-0484(2005) (Washington, DC: U.S. Government Printing Office, 2005).

2. EIA, *International Energy Outlook 2005*, p. 91. Quadrillion of British thermal units (Btu) is a standard unit that the Energy Information Administration uses to compare across energy sources.

3. United States Census Bureau, "Global Population Profile: 2002," (Washington, DC: U.S. Census Bureau, Population Division, 2004). Available at <http://www.census.gov/ipc/prod/wp02/wp-02.pdf>.

4. EIA, *Annual Energy Review 2004*, Report No. DOE/EIA-0384(2004) (Washington, DC: U.S. Government Printing Office, 2005), p. 137.

5. EIA, *International Energy Outlook 2005*, p. 93.

6. "Big Oil's Biggest Monster," *The Economist*, January 8, 2005, pp. 53–54.

7. Jad Mouawad, "Not a Ship to Spare: A Tanker Shortage Contributes to Rising Oil Costs," *New York Times,* October 20, 2004, p. C1.

8. "Gasoline (inflation adjusted)," online posting, June 10, 2004, retrieved January 27, 2005, from <http://bigpicture.typepad.com/comments/2004/06/gasoline_inflat.html>.

9. International Energy Agency (IEA), "World Energy Investment Outlook Sees Need for $16,000 Billion of Energy Investment through 2030, Highlights Major Challenges in Mobilizing Capital," Press release, November 4, 2003.

10. IEA, *Key World Energy Statistics 2005* (Paris: IEA, 2005). Available at <http://www.iea.org/bookshop/add.aspx?id=144>.

11. Jennifer Coleman, "Energy Crisis: California Lets High-Polluting Power Plants Resume," *Associated Press*, December 8, 2000.

12. EIA, *International Energy Outlook 2005*, p. 99.

13. "Global Power Plant Additions Reach Unprecedented Levels, According to Platts," April 5, 2004. Available at <http://www.platts.com/Resources/Press Room/2004/040504.xml>.

14. The estimates of cost of generation for various electricity generators are drawn from multiple sources. The PV estimates come from the forecast model in chapter 5, modified for a slightly higher cost of capital for utility customers versus residential customers. The cost range for coal and natural gas comes from John Deutch, Ernest Moniz, et al., *The Future of Nuclear Power: An MIT Interdisciplinary Study,* (Cambridge, MA: MIT, 2003). The rest use the estimates cited in chapter 4.

15. Janet Sawin, *Mainstreaming Renewable Energy in the Twenty-First Century* (Washington, DC: Worldwatch Institute, 2004), p. 13.

16. David Nichols and David Von Hippel, *Best Practices Guide: Integrated Resource Planning for Electricity* (Boston: Tellus Institute, 2000), p. 15.

17. Nichols and von Hippel, *Best Practices Guide*, p.15.

18. The estimates here include all of those used in figure 6.2 and a range of wholesale electricity prices for intermediate and peak loads from discussions with multiple industry professionals. The solar-thermal-electricity estimates use Wim C. Turkenburg (ed.), "Renewable Energy Technology," in *World Energy Assessment: Energy and the Challenge of Sustainability* (New York: United Nations Development Program [UNDP], 2000), p. 266, less estimates of 25 percent cost reductions in the technology since then.

19. Wolfgang Eichhammer et al., *Assessment of the World Bank / GEF Strategy for the Market Development of Concentrating Solar Thermal Power* (Washington, DC: Global Environment Facility, 2005). Available at <http://thegef.org/Documents/Council_Documents/GEF_C25/C.25.Inf.11_World_Bank_Assessment.pdf>, p. iv.

20. Tri-State Generation and Transmission Association, "Springerville's World-Class Solar Site," *Network*, Fall 2004. Available at <http://www.tristategt.org/NewsCenter/Magazine/Archives/NetworkFALL04.pdf>.

21. IEA, *Renewables for Power Generation: Status and Prospects 2003 Edition* (Paris: IEA, 2003). Available at <http://www.iea.org/textbase/nppdf/free/2000/renewpower_2003.pdf>.

22. Sanyo, "2002 Environmental Management Report," September 2002 (Osaka, Japan: Sanyo Electric Co., Ltd., 2002). Available at <http://www.sanyo.co.jp/environment/pdf/pdf_data/kan2002_e.pdf>.

Chapter 7

1. Statistics Bureau, Japan Ministry of Internal Affairs and Communications, *Statistical Handbook of Japan 2005* (Tokyo: Statistics Bureau, 2005). Available at <http://www.stat.go.jp/english/data/handbook/c07cont.htm>, chapter 2.

2. International Energy Agency (IEA), *Key World Energy Statistics 2005* (Paris: IEA, 2005). Available at <http://www.iea.org/bookshop/add.aspx?id=144>.

3. Paul Maycock, "Market Update: Global PV Production Continues to Increase," *Renewable Energy World* 8(4) (July–August 2005): 86–99, pp. 92, 96.

4. Energy Information Administration (EIA), "Electricity Prices for Households," available at <http://www.eia.doe.gov/emeu/international/ elecprih.html>.

5. EIA, *Electric Power Monthly* (October 2005): 2; EIA, *Electric Power Monthly* (October 2003): 85.

6. Jonathon Gatehouse, "Blackout Hits Ontario and Seven U.S. States," Maclean's Magazine (August 2003). Available at <http://www.thecanadianencyclopedia.com/index.cfm?PgNm=TCE&Params=M2SEC787505>.

7. Database of State Incentives for Renewable Energy (DSIRE), "Financial Incentives." Available at <http://www.dsireusa.org/summarytables/financial.cfm?&CurrentPageID=7>.

8. DSIRE, "Rules, Regulations and Policies." Available at <http://www.dsireusa.org/summarytables/reg1.cfm?&CurrentPageID=7>.

9. Niels I. Meyer, "Renewable Energy Policy in Denmark," *Energy for Sustainable Development* 8(1) (2004). Available at <http://www.ieiglobal.org/ESDVol8No1/05denmark.pdf>.

10. IEA, *Key World Energy Statistics 2005.*

11. EIA, "Table 5.6.B: Average Retail Price of Electricity to Ultimate Customers by End-Use Sector, by State, Year-to-Date through July 2005 and 2004," *Electric Power Monthly*, October 13, 2005, retrieved October 25, 2005, from <http://www.eia.doe.gov/cneaf/electricity/epm/table5_6_b.html>.

12. The figures in the graph come from the electricity prices referenced in the same paragraph and PV electricity cost estimates from the model in chapter 5.

13. Statistics Bureau, *Statistical Handbook of Japan 2005,* chapter 7.; Energy Information Administration (EIA), "Germany: Country Analysis Brief" (Washington, DC: U.S. Department of Energy, 2005). Available at <http://www.eia.doe.gov/emeu/cabs/germany.html>.

14. Jesse Broehl, "Washington State Passes Progressive Renewable Energy Legislation," *Renewable Energy Access* (Peterborough, N.H.: Renewable Energy Access, 2005). Available at <http://renewableenergyaccess.com/rea/news/story?id=34647>.

15. The isocost curves in this graph were developed from the model in chapter 5. Insolation came from NASA satellite data. Electricity estimates came from EIA, "Table 5.6.B," the same used in figure 7.2.

16. Amory Lovins, et al., *Small Is Profitable: The Hidden Economic Benefits of Making Electrical Resources the Right Size* (Snowmass, CO: Rocky Mountain Institute, 2002), p. 118.

17. Kate Martin, "Power by the Hour," *Photon International* (December 2004): 38.

Chapter 8

1. Lori Bird and Blair Sweezy, *Green Power Marketing in the United States: A Status Report* (6th ed.), Report No. NREL/TP-620-35119 (Golden, CO: National Renewable Energy Laboratory, 2003), p.1.

2. Lori Bird and Karen Cardinal, *Trends in Utility Pricing Programs*, Report No. NREL/TP-620-36833 (Golden, CO: National Renewable Energy Laboratory, 2004), p. 1.

3. British Petroleum, "Project Profiles: Whole Foods Market—USA." Available at <http://www.bp.com/genericarticle.do?categoryId=3050476&contentId=3060095>; Jesse Broehl, "Wal-Mart Deploys Solar, Wind, Sustainable Design," *Renewable Energy Access* (Peterborough, NH: Renewable Energy Access, 2005),

available at <http://renewableenergyaccess.com/rea/news/story?id=34647>; Solar Integrated Technologies, "Customers by Industry: Industrial and Commercial," available at <http://www.solarintegrated.com/CUS_IN_Com.html>.

4. Jim Carlton, "J. P. Morgan Adopts 'Green' Lending Policies," *Wall Street Journal*, April 25, 2005, p. B1.

5. PowerLight, "Products and Services." Available at <http://www. powerlight. com/products>.

6. United States Department of Energy (DOE), "Building America," retrieved January 27, 2005, from <http://www.eere.energy.gov/buildings/building_america/about.html>.

7. Travis Bradford, "Future of Solar Energy Policy in Industrial Markets: The Move towards Zero-Energy Homes," presented at the 2005 Solar World Conference in Orlando, FL, August 6–12, 2005 (Cambridge, MA: Prometheus Institute, 2005).

8. "Support Solar Power in California!" Planning Conservation League. Available at <http://action.nwf.org/pcl/alert-description.tcl?alert_id=892065>.

9. Real Goods Web site. Available at <http://www.realgoods.com>.

10. International Energy Association (IEA), *World Energy Outlook 2002* (Paris: OECD, 2001), chapter 13 (Energy and Poverty), p. 13.

11. IEA, *World Energy Outlook 2002*.

12. Eric Martinot, Anil Cabraal, and Subodh Mathur, "World Bank / GEF Solar Home Systems Projects: Experiences and Lessons Learned 1993–2000," *Renewable and Sustainable Energy Reviews* 5(1) (2001): 39–57. Available at <http://www.gefweb.org/ResultsandImpact/Experience_and_Lessons/WBGEF_SHS_RSER.pdf>.

13. Elizabeth W. Cecelski, "From Rio to Beijing: Engendering the Energy Debate," *Energy Policy* 23(6): 567.

14. IEA, "Financing Mechanisms for Solar Home Systems in Developing Countries: The Role of Financing in the Dissemination Process," Report IEA-PVPS T9-01:2002 (Paris: IEA, 2002). Available at <http://www.oja-services .nl/iea-pvps/products/download/rep9_01.pdf>.

15. Grameen Web site. Available at <http://grameen-info.org/grameen/gshakti/programs.html>.

16. Martinot et al., "World Bank/GEF Solar Home Systems Projects," pp. 4–5.

17. Daniel M. Kammen, Kamal Kapadia and Matthias Fripp, "Putting Renewables to Work: How Many Jobs Can the Clean Energy Industry Generate?" RAEL Report, (Berkeley, CA: University of California, 2004), p. 1. Kammen compares the various generation technologies by derating them to equivalent expected output of electricity.

18. ITN Energy Systems Inc, "Photovoltaic Reverse Osmosis Desalination System," (Washington, DC: U.S. Department of the Interior, 2004), available at <http://www.usbr.gov/pmts/water/media/pdfs/report104.pdf>; M. Thomson

et al., "Batteryless Photovoltaic Reverse-Osmosis Desalination System," U.K. Department of Trade and Industry, 2001, available at <http://www.dti.gov.uk/energy/renewables/publications/pdfs/SP200305.pdf>.

19. Smithsonian Institution, "Growing Pains," *Ocean Plant* (Traveling Exhibit), 1995, retrieved January 18, 2006, from <http://seawifs.gsfc.nasa.gov/OCEAN_PLANET/HTML/peril_population. html>.

20. Kevin Wilson, "Electric Car," *Microsoft Encarta Online Encyclopedia 2005*. Available at <http://encarta.msn.com/encyclopedia_761580732_2/Electric_Car.html>.

Chapter 9

1. Norman Myers and Jennifer Kent, *Perverse Subsidies: How Tax Dollars Can Undercut the Environment and the Economy* (Washington, DC: Island Press, 2001).

2. Robert Jay Dilger, *American Transportation Policy* (Westport, CT: Praeger, 2003).

3. Myers and Kent, *Perverse Subsidies*, p. 85.

4. Myers and Kent, *Perverse Subsidies*, p. 85.

5. Myers and Kent, *Perverse Subsidies*, p. 87.

6. Myers and Kent, *Perverse Subsidies*, p. 12.

7. Environmental Protection Agency (EPA), "Allowance Trading Basics," EPA Clean Air Markets (2005). Available at <http://www.epa.gov/airmarkets/trading/basics/>.

8. "Revving Up," *The Economist*, July 7, 2005, pp. 64–65.

9. Steve Raabe, "National Renewable Energy Laboratory Gets New Director in Golden, Colorado," Denver Post, January 6, 2005. Available at <http://www.enn.com/today.html?id=6881>.

10. United States Department of Energy (DOE), "Energy Research at DOE: Was It Worth It? Energy Efficiency and Fossil Energy Research 1978–2000" (Washington, DC: DOE, Committee on Engineering and Technical Systems, 2001). Available at <http://books.nap.edu/catalog/10165.html>.

11. Robert Margolis and Daniel Kammen, "Underinvestment: The Energy Technology and R&D Policy Challenge," *Science Magazine* (July 20, 1999). Available at <http://socrates.berkeley.edu/~rael/Margolis&Kammen-Science-R&D.pdf>.

12. Travis Bradford, "Future of Solar Energy Policy in Industrial Markets: The Move towards Zero-Energy Homes," presented at the 2005 Solar World Conference in Orlando, FL, August 6–12, 2005 (Cambridge, MA: Prometheus Institute, 2005).

13. New Energy Foundation (NEF), "New and Renewable Energy in Japan." Available at <http://www.nef.or.jp/english/new/index.html>.

14. International Energy Agency (IEA), *Key World Energy Statistics 2005* (Paris: IEA, 2005). Available at <http://www.iea.org/bookshop/add.aspx?id=144>.

15. Jackie Jones, "Japan's PV Market," *Renewable Energy World* (March–April 2005): 36.

16. For the period 2000 through 2004. Solarbuzz Inc, *MarketBuzz 2005: Annual World PV Market Review*, (San Francisco: Solarbuzz Inc., 2005), p.15.

17. Paul Maycock, "PV Market Update: Global PV Production Continues to Increase," *Renewable Energy World* 8(4) (July–August 2005): 86–99, p. 95.

18. European Renewable Energy Council (EREC), "Renewable Energy Policy Review: Germany" (Brussels: EREC, 2004). Available at <http://www.erec-renewables.org/documents/RES_in_EUandCC/Policy_reviews/EU_15/Germany_policy_final.pdf>.

19. Christoph Hünnekes, "Germany: Photovoltaic Technology Prospects and Status," International Energy Agency, 2004. Available at <http://www.oja-services.nl/iea-pvps/ar03/deu.htm>.

20. Solarbuzz, *Market Buzz 2005*, p. 15.

21. "Germany Renounces Nuclear Power," June 15, 2000, *BBC News*. Available at <http://news.bbc.co.uk/1/hi/world/europe/791597.stm>.

22. Anne Kreutzmann, "Complaints at the Highest Level," *Photon International* (January 2005): 17.

23. Michael Schmela, "Green Light for Solar Power: China's PV Industry Awakens," *Photon International* (September 2005): 56.

24. EREC, "Renewable Energy Policy Review: Spain" (Brussels: EREC, 2004). Available at <http://www.erec-renewables.org/documents/RES_in_EUandCC/Policy_reviews/EU_15/Germany_policy_final.pdf>.

25. Jesse Broehl, "Solar Granted a Major Victory in Energy Bill," July 29, 2005, *Renewable Energy Access*. Available at <http://www.renewableenergyaccess.com/rea/news/story?id=34850>.

26. "Residential Programs: Solar Incentives," Los Angeles Department of Water and Power (LADWP), retrieved February 3, 2005, from <http://www.ladwp.com/ladwp/cms/ladwp000787.jsp>, also available at <http://www.dsireusa.org/library/includes/incentive2.cfm?Incentive_Code=CA15F&state=CA&CurrentPageID=1>.

27. Database of State Incentives for Renewable Energy (DSIRE), "Rules, Regulations and Policies." Available at <http://www.dsireusa.org/summarytables/reg1.cfm?&CurrentPageID=7>.

28. North American Board of Certified Energy Practitioners (NABCEP), "PV Installer Certification." Available at <http://www.nabcep.org/pv_installer.cfm>.

29. Gary Rivlin, "Green Tinge Is Attracting Seed Money to Ventures," *New York Times,* June 21, 2005, p. C1.

30. Anne Kreutzmann, "Rush to Trading Floor: PV Companies Flock towards the Stock Market," *Photon International* (October 2005): 52.

Chapter 10

1. Martin Green, "Third-Generation Solar: Future Photovoltaics," *Renewable Energy World* (July–August 2004): 203.

2. Sonnet L'Abbe, "Nanotechnologists' New Plastic Can See in the Dark," January 10 2005, *News@UofT*. Available at <http://www.news.utoronto.ca/bin6/050110-832.asp>.

3. Donald W. Aitken, "The Renewable Energy Transition: Can it Really Happen?" *Solar Today* (January/February 2005): 16–19.

4. U.S. Department of Energy (DOE), "Technology FactSheet: Photovoltaics at a Glance," retrieved August 19, 2004, from <http://www.eere.energy.gov/state_energy/technology_factsheet.cfm?techid=1>.

5. National Renewable Energy Laboratory (NREL), "An Information Resource for PV: Frequently Asked Questions about Photovoltaics (PV)." Available at <http://www.nrel.gov/ncpv/pvmenu.cgi?site+ncpv&idx=3&body=faq.html>.

Suggested Readings

Diamond, Jared. *Collapse: How Societies Choose to Fail or Succeed.* New York: Viking Penguin, 2005.

Goodstein, David. *Out of Gas: The End of the Age of Oil.* New York: Norton, 2004.

Goswami, D. Yogi, Frank Kreith, and Jan F. Kreider. *Principles of Solar Engineering.* 2nd ed. Philadelphia: Taylor & Francis, 2000.

Grubler, Arnulf, Nebojsa Nakicenovic, and William D. Nordhaus (Eds.). *Technological Change and the Environment.* Washington, DC: RFF Press, 2002.

Kuhn, Thomas S. *The Structure of Scientific Revolutions.* 3rd ed. Chicago: University of Chicago Press, 1996.

Perlin, John. *A Forest Journey: The Story of Wood and Civilization.* Woodstock, VT: Countryman Press, 2005.

Perlin, John. *From Space to Earth: The Story of Solar Electricity.* Cambridge, MA: Harvard University Press. 2000.

Renner, Michael, and Molly O. Sheehan for the Worldwatch Institute. *Vital Signs 2003: The Trends That Are Shaping Our Future.* New York: Norton, 2003.

Scheer, Hermann. *The Solar Economy: Renewable Energy for a Sustainable Global Future.* London: Earthscan, 2002.

Scheer, Hermann. *A Solar Manifesto.* 2nd ed. London: James and James, 2001.

Stefik, Mark, and Barbara Stefik. *Breakthrough: Stories and Strategies of Radical Innovation.* Cambridge, MA: MIT Press, 2004.

Tester, Jefferson W., Elisabeth M. Drake, Michael J. Driscoll, Michael W. Golay, and William A. Peters. *Sustainable Energy: Choosing Among Options.* Cambridge, MA: MIT Press, 2005.

Victor, David G. *The Collapse of the Kyoto Protocol and the Struggle to Slow Global Warming.* Princeton: Princeton University Press, 2001.

Credits

Figures

Figure 2.1
Nebojsa Nakicenovic and Arnulf Grübler, "Energy and the Protection of the Atmosphere," *International Journal of Global Energy Issues,* 13, nos. 1–3 (2000), 4-57, Reprinted as RR-00-18, International Institute for Applied Systems Analysis, Laxenburg, Austria.

Figure 2.2
Energy Information Administration (EIA), *Annual Energy Review 2004,* Report No. DOE/EIA-0384(2004) (Washington, D.C.: U.S.Government Printing Office). Available at <http://www.eia.doe.gov/emeu/aer/pecss_diagram.html>.

Figure 2.3
International Energy Agency (IEA), *Key World Energy Statistics 2005* (Paris: IEA, 2005). Available at <http://www.iea.org/bookshop/add.aspx?id=144>.

Figure 3.1
Arctic Climate Impact Assessment (ACIA), *Impacts of a Warming Arctic: Arctic Climate Impact Assessment* (Cambridge, UK: Cambridge University Press, 2004).

Figure 3.2
Association for the Study of Peak Oil and Gas (ASPO), *Newsletter* no. 58 (October 2005). Available at <http://www.peakoil.ie/downloads/newsletters/newsletter58_200> (accessed October 19, 2005).

Figures 5.2 and 5.3
Solarbuzz Inc., *MarketBuzz 2005: Annual World PV Market Review* (San Francisco: Solarbuzz Inc., 2005).

Figure 5.4
SolarPlaza.com, "The Spanish Solar PV Supply Chain," online posting, February 2005. Available at <http://www.solarplaza.com/content/pagina/2005/february%20news/43092> (retrieved October 28, 2005).

Figure 6.1
Energy Information Administration (EIA), *International Energy Outlook 2004,* Report No. DOE/EIA-0484 (2004) (Washington, DC: U.S. Government Printing Office, 2004). Available at <http://www.eia.doe.gov/oiaf/archive/ieo04/pdf/0484(2004).pdf>.

Figure 6.2
Coal and natural gas estimates John Deutch and Ernest J. Moniz, et al., *The Future of Nuclear Power: An Interdisciplinary MIT Study* (Cambridge, MA: MIT). Available at <http://www.eere.energy.gov/windandhydro/hydro_potential.html>.

Wind estimates American Wind Energy Association (AWEA), "The Economics of Wind Energy" (Washington, DC: AWEA, 2002). Available at <http:// www.awea.org/ pubs/factsheets/EconomicsOfWind-March2002.pdf>.

Hydro estimates United Nations Development Program (UNDP), *World Energy Assessment: Energy and the Challenge of Sustainability* (New York: UNDP, 2000). Available at <http://www.undp.org/seed/eap/activities/wea/index.html>.

Nuclear estimates John Deutch and Ernest J. Moniz, et al., *The Future of Nuclear Power,* and Amory B. Lovins and L. Hunter Lovins, "The Nuclear Option Revisited," *Los Angeles Times*, July 8, 2001. Available at <http://www.rmi.org/images/other/Energy/E01-19_NuclearOption.pdf>.

Geothermal estimates World Bank, "Geothermal Energy" (New York: World Bank). Available at <http://www.worldbank.org/html/fpd/energy/geothermal/>.

Figure 6.3
Public Service Commission of Wisconsin (PSC) and Wisconsin Department of Natural Resources (DNR), *Fox Energy Generation Project: Final Environmental Impact Statement,* Docket No. 05-CE-115 (Madison, WI: PSC, August 2002), p. 12.

Figure 6.4
Coal and natural gas estimates John Deutch and Ernest J. Moniz, et al., *The Future of Nuclear Energy.*

Wind estimates American Wind Energy Association (AWEA), "The Economics of Wind Energy."

Hydro estimates United Nations Development Program (UNDP), *World Energy Assessment.*

Nuclear estimates John Deutch and Ernest J. Moniz, et al., *The Future of Nuclear Power,* and Amory B. Lovins and L. Hunter Lovins, "The Nuclear Option Revisited."

Geothermal estimates World Bank, "Geothermal Energy."

Solar estimates Travis Bradford.

Figure 6.5
Winfried Hoffmann, et al., "Towards an Effective European Industrial Policy for PV Solar Electricity," presented at the 19th European Photovoltaic Solar Energy Conference and Exhibition, Paris, June 10, 2004. Available at <http://www.epia.org/04events/docs/Paris_Hoffman_IndustryPolicy.pdf>.

Figure 7.1
National Renewable Energy Laboratory (NREL), "U.S. Solar Radiation Resource Maps: Annual Flat Plate South at Latitude." Available at <http://rredc.nrel.gov/solar/old_data/nsrdb/redbook/atlas/>.

Figure 7.2
International Energy Agency (IEA), *Key World Energy Statistics 2005* (Paris: IEA, 2005), available at <http://www.iea.org/bookshop/add.aspx?id=144>; Energy Information Administration (EIA), "Table 5.6.B: Average Retail Price of Electricity to Ultimate Customers by End-Use Sector, by State, Year-to-date through July 2005 and 2004," *Electric Power Monthly* (October 13, 2005), available at <http://www.eia.doe.gov/cneaf/electricity/epm/table5_6_b.html> (accessed October 25, 2005).

Figure 7.3
Energy Information Administration (EIA), "Table 5.6.B: Average Retail Price of Electricity to Ultimate Customers by End-Use Sector, by State, Year-to-date through July 2005 and 2004"; Apricus Solar Co., Ltd, "Insolation: USA," available at <http://www.apricus-solar.com/index.htm>, citing the NASA Surface Meteorology and Solar Energy Data Set (available at <http://eosweb.larc.nasa.gov/cgi-bin/sse/grid.cgi?uid=3030>).

Figure 9.1
William P. Hirshman, "Small Talk and Big Plans in Japan," *Photon International* (May 2004): 65–67.

Figure 9.2
Database of State Incentives for Renewable Energy (DSIRE), "Rebate Programs for Renewable Energy Technology," October 2005. Available at <http://www.dsireusa.org/library/includes/topic.cfm?TopicCategoryID=6&CurrentPageID=10>.

Table

Table 5.1
Paul Maycock, "Market Update: Global PV Production Continues to Increase," *Renewable Energy World* 8, no. 4 (July–August 2005): 86–99.

Index